For EU product safety concerns:
IN TECH d.o.o., Prolaz Marije Krucifikse Kozulić 3, 51000 Rijeka, Croatia,
info@intechopen.com or visit our website at intechopen.com.

Meet the editor

Dr. Aydin received a BS and MS in Engineering Physics from Ankara University, Turkey, in 1989 and 1993, respectively. He also obtained a Ph.D. in Physical Chemistry from the City University of New York (CUNY), USA, in 2001. From 2001 to 2005, he served as a post-doctoral associate at the Center for Analysis of Structures and Interfaces, Chemistry Department, The City College of New York, USA. In 2008, Dr. Aydin joined the faculty of the Chemistry Department, Ondokuz Mayıs University, Turkey, where he rose through the ranks from assistant professor to Professor of Physical Chemistry. His works have been published in numerous respected professional journals and he has given numerous talks at both national and international conferences.

Contents

Preface

Boron is a chemical element with three valence electrons that can form many new compounds with unique physical, chemical, and electrical properties by incorporating them with other atoms. Boron is a semimetal that is found in low amounts both in the solar system and in the Earth's crust. It is produced by cosmic rays and high-energy radiation from supernovae that impact boron-containing minerals. Naturally occurring boron is found within chemical compounds such as borate minerals; however, the elemental form is not found because of its high reactivity. Also, high-purity boron is difficult to produce industrially because of contamination by carbon and other elements at high temperatures. Boron has several allotropes, including amorphous boron, a dark brown powder, and crystalline boron, a black material with a melting point above 2000°C, which is extremely hard and weakly conductive at room temperature but a good conductor at high temperatures.

Boron, in synthetic chemistry and research chemicals, is a significant component of reducing agents such as sodium borohydride and borane. In the Corey–Bakshi–Shitaba catalyst used to asymmetrically reduce ketones, boron plays dual roles as a hydride source and a Lewis acid. Boron's Lewis acid properties result from an empty p-orbital, which confers electron-accepting abilities, such as in the familiar boron trifluoride used as a Lewis acid catalyst. Boronic acids and esters are key organic building blocks in synthetic chemistry. These oxygen- and carbon-containing boron compounds are important cross-coupling partners in palladium-catalyzed Suzuki–Miyaura reactions. They also play a critical role in various fields, including small molecule synthesis of pharmaceuticals, agrochemicals, biomedical sciences, material science, electronic devices, nuclear industry, and fire retardants.

The advancements in particle beam therapy have led to the development of treatment methods (such as boron neutron capture therapy, proton-boron capture therapy, and light ion beam therapy), which show promise in the treatment of malignant tumors. Boron and its compounds are utilized in these therapies and have diverse applications as anticancer agents, antimicrobial agents, and components of imaging agents for optical and nuclear imaging. These developments contribute to the progress of medical science and offer new possibilities for effective and targeted treatment approaches. Additionally, boron and boron compounds find applications in radiation-shielding materials. Due to their excellent neutron-absorption properties, boron-based materials are utilized in the construction of shielding materials for radiation protection. They help attenuate and absorb neutron radiation, reducing the potential risks associated with radiation exposure. Furthermore, the rapid development of nanoscience and nanotechnology has inspired scientists to investigate new features of boron nanostructures such as boron-containing nanotubes, nanowires, nanosheets, and nanocomposites, among others. These novel structures have found applications in energy conversion, hydrogen storage, batteries, catalysts, capping agents, electronics, superconductors, mechanically strong components, agriculture, nanomedicine, cancer research, and much more.

The book contains seven chapters, each focusing on the design and development of relevant boron compounds and their corresponding applications in diverse fields. The chapters are organized into two sections.

Section 1, "Role and Applications of Boron and Its Compounds in Biomedicine, Health and Agriculture" includes three chapters. Chapter 1, "Physical, Biological, and Clinical Merits of High Energy Boron Ions for Radiation Therapy", explores the advantages and applications of high-energy boron ions in radiation therapy, particularly for cancer treatment. It discusses the physical properties of boron ions, their biological effects on cancer cells, and the potential clinical benefits of targeted radiation therapy. Chapter 2, "Boron Removal by Donnan Dialysis According Doehlert Experimental Design", examines the use of Donnan dialysis and the Doehlert experimental design to remove boron from solutions. The focus is on optimizing the dialysis process parameters for efficient boron removal, which may have implications for industrial processes or wastewater treatment. Chapter 3, "Boron Nutrition in Horticultural Crops: Constraint Diagnosis and Their Management", discusses the importance of boron in the nutrition of horticultural crops, such as fruits and vegetables. It may cover diagnosing boron deficiencies or toxicities in these crops and strategies for managing these issues to ensure optimal growth and yield.

Section 2, "Role and Applications of Boron and Its Compounds in Industrial and Nanotechnology Fields", includes four chapters. Chapter 4, "Boron and Boron Compounds in Radiation Shielding Materials", explores the use of boron and boron-based compounds in the development of radiation-shielding materials. These materials are essential for protecting against radiation in various applications, including nuclear power plants and medical facilities. Chapter 5, "Some Features of Boron Isotopes Separation by Laser-Assisted Retardation of Condensation Method", discusses the separation of boron isotopes using the laser-assisted retardation of condensation method. Isotope separation is crucial for various applications, and the use of lasers in the process may offer advantages in terms of precision and efficiency. Chapter 6, "Effect of Capping Agents on the Nanoscale Metal Borate Synthesis", examines the influence of capping agents on the synthesis of nanoscale metal borates. Capping agents play a critical role in controlling the size and properties of nanoparticles during their formation, making them important for applications in catalysis and nanocomposite materials. Finally, Chapter 7, "Numerical Simulation of the Mechanical Behaviour of Boron Nitride Nanosheets and Nanotubes", discusses using numerical simulations, such as Nanoscale Continuum Modelling (NCM) or Molecular Structural Mechanics (MSM) approaches, to investigate the mechanical properties of boron nitride nanosheets and nanotubes. Understanding their mechanical behavior is crucial for potential applications in nanotechnology, electronics, and composite materials.

This book offers readers a comprehensive exploration of the developments and applications of boron and boron compounds in various fields. It is a useful resource for researchers, scientists, engineers, and students interested in these fields, providing them with essential insights and knowledge to advance their work and studies.

I would like to express my deep appreciation to all the contributing authors who are experts in their research fields: Brahme Anders, Marzouk Trifi Ikhlass, Lasaad Dammak, Lassaad Baklouti, Bechair Hamrouni, Pauline Alila, Ahmet Hakan Yilmaz,

Bülend Ortaç, Sevil Savaskan Yilmaz, Lyakhov Konstantin, Senberber Dumanli Fatma Tugce, Nataliya A. Sakharova, Jorge M. Antunes, Andre F. G. Pereira, Bruno M. Chaparro, and Jose V. Fernandes. I would also like to thank the staff at IntechOpen for their constant support throughout the planning and compilation of this book. Finally, I would like to express my sincere gratitude to my beloved wife Oya Aydın for her unflinching understanding and patience.

Metin Aydin
Faculty of Science,
Department of Chemistry,
Ondokuz Mayıs University,
Samsun, Turkey

Section 1

Role and Applications of Boron and Its Compounds in Biomedicine, Health and Agriculture

Chapter 1

Physical, Biological, and Clinical Merits of High Energy Boron Ions for Radiation Therapy

Anders Brahme

Abstract

The lightest ions beyond protons, principally helium, lithium, and boron ions, make highly specific molecular Bragg peak radiation therapy of malignant tumors possible with minimal adverse normal tissue reactions. The Bragg peak ionization density is mainly elevated in a few mm wide spot at the end of these ions with substantially increased local apoptosis and senescence induction. Mainly placing Bragg peaks in the gross tumor volume with increased local therapeutic effect and only low ionization density and easily repairable damage in normal tissues. The possible geometrical accuracy of the dose delivery will be ≈1 mm with these ions. Interestingly, high-resolution molecular tumor imaging will then be possible, particularly with ^{8}Boron ions that are our lightest positron emitter allowing immediate accurate PET-CT imaging to delineate the target volume dose delivery. Compared to carbon ions the boron radiation damage to normal tissues in front of and behind the tumor is reduced at the same time as tumor apoptosis and senescence are increased. A mean tumor cure as high as 80% should be possible with Boron ion therapy using new clinical fractionation principles and even more when early tumor detection and malignancy estimation methods are brought into more regular clinical use.

Keywords: boron ion radiation therapy, $^{8-11}$boron ions, low dose apoptosis, low dose hypersensitivity, high dose apoptosis, radiation dose-response relationships, light ion radiation therapy, radiation therapy optimization

1. Introduction

The interaction of ionizing radiation with living tissues and tumors is one of the most complex biomedical problems, since it requires knowledge about atomic and nuclear physics and the generated secondary electrons, as well as the molecular biology dynamics of living tissues and cells and their complex damage repair systems [1]. Understanding radiation-induced cellular damage and repair is the key to optimal safety in the therapeutic and the diagnostic use of high-quality radiation beams. Biologically optimized intensity-modulated photons, electrons, and light ions represent the ultimate development of radiation therapy as shown in **Figure 1**. The absorbed dose and biological effect on normal tissues can be designed so it is as low as possible from a radiation physical point of view, at the same time, as the therapeutic

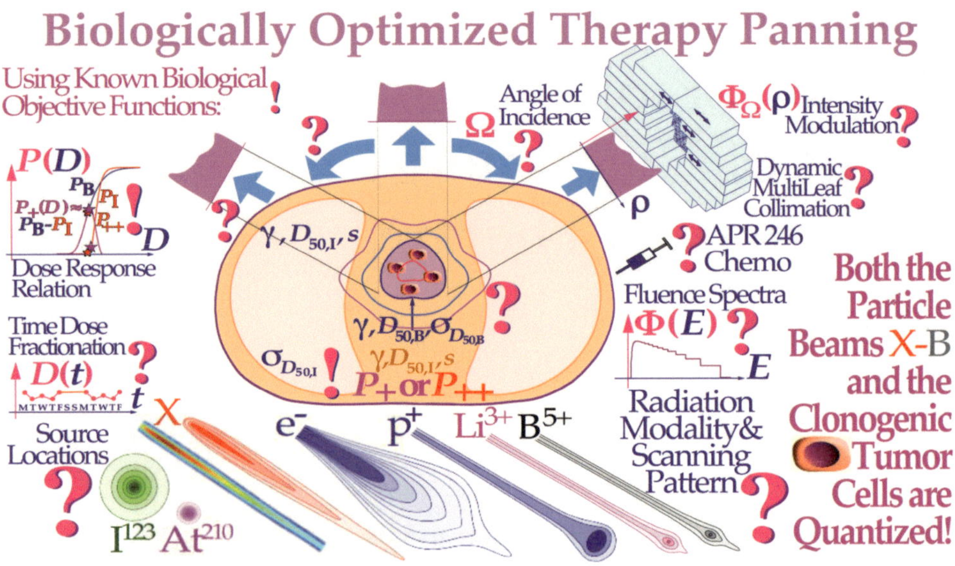

Figure 1.
Illustration of the fantastic power available by using biologically optimized inverse radiation therapy planning [2]. If we know the approximate sensitivity of the tumor and the normal tissues (!), it is possible to derive the biologically optimal beam directions and their intensity modulation (?; [3]) and it is even possible to find the optimal combination of low and high ionization density radiations (cf. **Figure 15**) *and their incident energy spectra as well as the ideal time dose fractionation [1, 4, 5] using biological complication free cure (P+) optimization strategies (P++: P+ with concomitant injury minimization [3]). In addition, if we have information about the interaction of the radiation modality of interest with chemotherapeutic agents of preference, the combined treatment schedule can also be optimized in biological terms. During the last week of treatment, only a hand full of tumor clonogens remain, and should thus preferably be treated with more microscopically uniform electron or photon beams. This is optimal since both the particle beam and the tumor cells are quantized and may protect some tumor clonogens from lethal hits by inevitable cold spots between the ions during the last most curative therapeutic dose fraction [1–5]!.*

effect on radiation-resistant tumor cells is as high as possible from a radiation biological point of view [2]. With the lightest ions above protons: He, Li, B, and C the border region between the clinical gross tumor and target volume and surrounding healthy normal tissues can be set as narrow as physically possible. In addition, the optimal number of treatment fractions can be substantially reduced, and the curative gain factor on radiation resistant hypoxic tumor cells may generally be more than doubled compared to low ionization density photons, electrons, and protons. **Figure 1** shows how the optimal selection of Therapeutic beams can be arranged and the energy modulation shaped to maximize the cure probability of the patient with minimal risk for side effects in normal tissues [2]. Largely based on clinically established dose-response parameters of normal tissues (γ, D_{50}, s).

2. Radiation biology of radiation therapy

2.1 Handling of DNA damage by the TP53 gene

The new interaction cross-section based Repairable-Homologous-Repairable damage formula for radiation-induced cellular inactivation, repair, misrepair, and apoptosis in TP53 intact and mutant cell lines can be used to optimize radiation therapy. The formulation requires renewed thinking about the biological optimization of radiation therapy. It suggests that most TP53 intact normal tissues are Low Dose Hyper Sensitive (LDHS, see left insert of **Figure 2**) and that the inherent microscopic heterogeneity of higher Linear Energy Transfer (*LET*) ion treatments the last week would benefit from low *LET* as shown in the lower right of **Figure 2** [1, 4, 5]. The ability of the new method to quantify apoptosis [1], has helped identify the early Low Dose Hypersensitivity (LDHS) and Low Dose Apoptosis (LDA) of most normal tissues and tumors with intact TP53 and ATM genes. This mechanism has probably been developed by nature's proses of survival advantage selection, to ensure minimal risk for severe mutations to the genome before the DNA repair system is fully functional after around a dose of 1Gy [1, 4–8]. As a compensating measure the apoptosis-inducing caspase 3 gene product (**Figure 2** right) remarkably "remembers" this low dose apoptotic cell loss and starts cellular repopulation to reestablish homeostasis in the tissues after being irradiated.

This useful mechanism in normal tissue is a well-known problem after suboptimal radiation therapy where it can cause accelerated repopulation of remaining tumor cells at the end of treatment [9]. A clear curative intent is probably the principal way to avoid this tumor-reactivating mechanism. These studies also identified that maximum apoptosis is induced by the lowest *LET* ions largely as they have the highest fluence of δ-electron apoptosis induced by primary ions per unit dose [1, 4]. With a too high *LET* the apoptosis and senescence will instead be high in the normal tissues in front of and behind the tumor, which may be undesirable from a complication-free

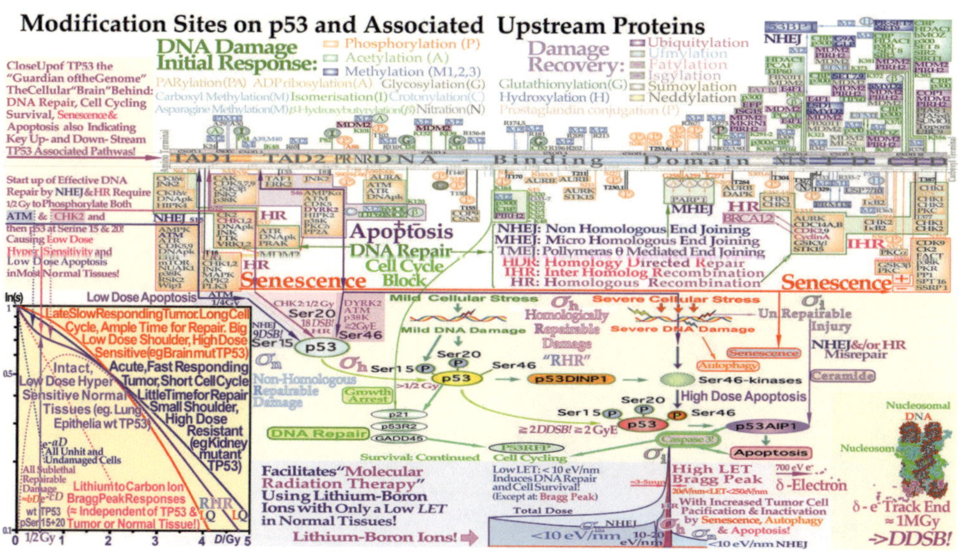

Figure 2.
The complex responses of the TP53 gene to mild and severe genetic stress is largely determining the cellular response to radiation [1–13]. Mild stress phosphorylates the serine 15 and 20 sites on p53 by ATM and CHK2, resulting in cell cycle block and DNA repair. This results in LDHS in normal tissues but generally not in tumors often with a mutant TP53 gene as seen in the cell survival insert. Local high doses or high ionization densities resulting in DDSBs (Dual Double Strand Breaks cf. **Figure 3***) [2, 14] are increasing the severity of the damage phosphorylating also the serine 46 site e.g., via DYRK2, p38K, or ATM, and a high dose apoptotic (HDA) response may get triggered. Lithium-Boron (cf* **Figure 12***) ions allow unique therapeutic use by inducing a massive apoptotic-senescent tumor cell response within the Bragg peak (σ_h homologically repairable damage and σ_i direct inactivation cross-sections cf.* **Figure 4** *[1, 4]), but in front of and beyond the Bragg peak, the LET is low, and non-homological easily repairable damage is mainly induced (σ_n cross-section see* **Figure 4** *insert [1: Figure 8, 4, 7, 8]).*

cure point of view, even if hypoxic tumors may marginally benefit from a high *LET* (cf [4] and **Figure 3**).

2.2 Dual double strand breaks

The severe Bragg peak damage to supercoiled DNA first wound two times around nucleosomes in the cell nucleus is shown in the lower right corner of **Figure 2** and the close-up in **Figure 3**. Most toxic are the ≈700 eV electrons that deposit a dose in the neighborhood of the track of up to 10^6 Gy as seen in the right of **Figure 2** [1, 4, 14, 15]. In front of the Bragg peak, the density of electrons is lower and even more so in the high energy entrance region requiring many more ions to deliver a dose of around 2Gy (cf **Figure 7**). This phenomenon explains why the medium and low *LET* ion beams are most efficient in inducing apoptosis (as further seen in **Figures 11** and **13**) and thereby eradicating hypoxic tumor cells. With a low *LET*, too few severe direct cell kill events are obtained and at high *LET* too few ions are available at a given dose even though they produce very severe damage. The most probable DNA fragment length at high doses is around 78 base pairs, corresponding to a single turn around the nucleosome (cf **Figure 3**). A δ-electron track end that may randomly hit the DNA at any point on the periphery of the nucleosome and then often produce a Dual DSB (DDSB), can easily produce such fragments. This will very often make DNA fragments of close to a single nucleosomal DNA turn in length as seen in **Figure 3**. This string of DNA may easily be lost as a micronuclei or gey inserted erroneously to make a severe mutation and possibly a non-functional protein. Fortunately, most simple DSBs are repaired

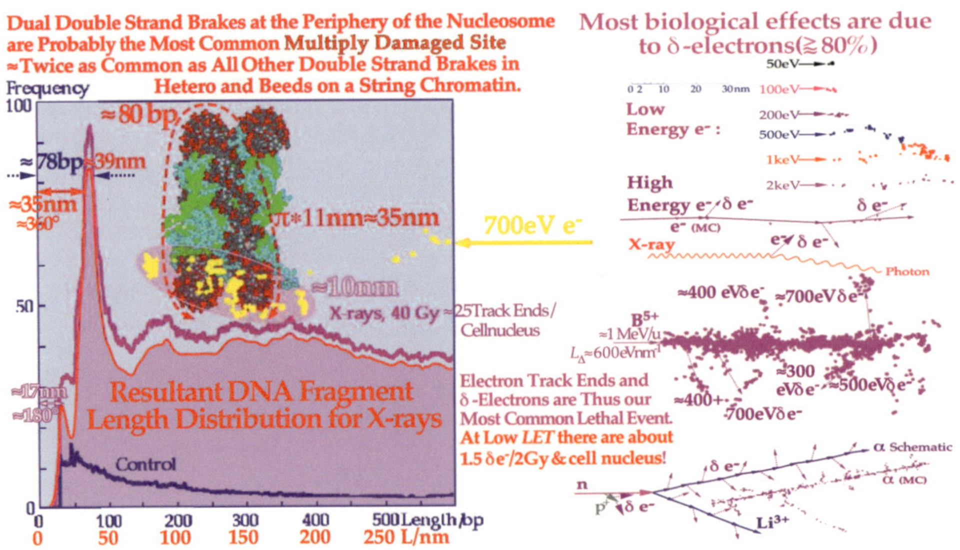

Figure 3.
Molecular close-up of ion tracks (right) showing that most of the lethal cell damage of densely ionizing ions is induced by low energy δ-electrons in the 200 eV to 1 keV energy range generating severe difficult-to-repair DNA damage in the cell nucleus such as dual DSBs at the periphery of the nucleosome (left, cf. [14]). The left insert shows that the most common DNA segment length corresponds to a single turn of DNA around a nucleosome as expected from the DDSBs at the periphery of a nucleosome. Interestingly, the 78–80 base pair fragments are about twice as common as all other fragment sizes and they should be expected to be even more common with high LET beams having ≈3 times more secondary δ-electrons in the sub-keV energy range, with a very high probability of inducing lethal DDSBs.

correctly (>99%) so a DDSB is really the most common multiply damaged site causing severe cell loss [1, 4, 14, 16].

2.3 Cell survival

In two recent DNA repair-based publications [1, 4], the accurate quantification of the cellular survival and damage to tumors and normal tissues are developed to significantly improve our ability to precisely describe the survival to low and high ionization density (*LET*) radiations and doses $S = e^{-aD} + bDe^{-cD}$, far beyond the possibilities of the conventional linear quadratic cell survival model ($S \approx e^{-\alpha D-\beta D^2}$). Not only are the undamaged cells (e^{-aD}) separated from the sublethal damaged cells (*bD*) but also the two major DNA damage repair pathways, namely homologous and non-homologous end joining DNA repair (HR, NHEJ) can each be identified ($b_h De^{-chD} + b_n De^{-cnD} \approx bDe^{-cD}$) and so can their complex interactions cf. **Figures 4** and 5 [1].

It is therefore given the name the repairable-homologous-repairable or RHR-formulation as seen in the left insert in **Figures 2** and **4** (cf [1, 4] for further details). The fractionation window linked to LDHS normal tissues as seen in the left insert in **Figure 2** indicates that the low *LET* dose to organs at risk should be ≈2 Gy/Fr as this produces the least damage per unit dose, whereas the tumor dose should be substantially higher to ensure perfect tumor cure [5]. This calls for biologically optimized treatments using a few intensity-modulated beams [3] to avoid secondary cancer risk and get a true curative intent, avoiding caspase 3 induced accelerated tumor cell repopulation (see right part of the middle half of **Figure 2**, [9]). Light ions with the lowest possible *LET* in normal tissues and high *LET* only in the tumor indicate lithium to boron ions [5]. The high microscopic heterogeneity in the tumor will cause local microscopic cold spots. Therefore, the last week of curative ion therapy, with few remaining viable tumor clonogens randomly spread in the target volume, as indicated in **Figure 1**, should receive the last 10 GyEquivalent by low *LET* ensuring perfect microscopic tumor coverage and high cure and reduced risk for adverse reactions in normal tissues [5]. Interestingly, such an approach would also ensure a steeper rice of tumor cure and a higher complication-free cure as few remaining clonogens are fairly well oxygenated eliminating shallower tumor response by ion heterogeneity (see **Figure 17**). Avoiding ion microscopic heterogeneity in normal tissues increases complication-free cure both at the low dose normal tissue complication side and high dose tumor cure end of the treatment [5].

2.4 Apoptosis induction

The major forms of interaction between homologous and non-homologous DNA repair such as homologous repair of non-homologous misrepair are accounted for and the probability to induce programmed cell death (Apoptosis) and potentially even more so, permanent cell cycle arrest (Senescence) as well as other cell cycle losses [3, 4]. Interestingly, these processes are probably the optimal ways to inactivate a tumor with minimal inflammatory response and without massive immediate tumor decomposition. Interestingly, lithium ions and its neighbors helium, beryllium, and boron ions have an important and unique potential to induce apoptosis locally, mainly in a few mm-size volumes around their deep high ionization density Bragg peaks in the tumor as seen in **Figures 4, 5** and 7 (Be may be associated

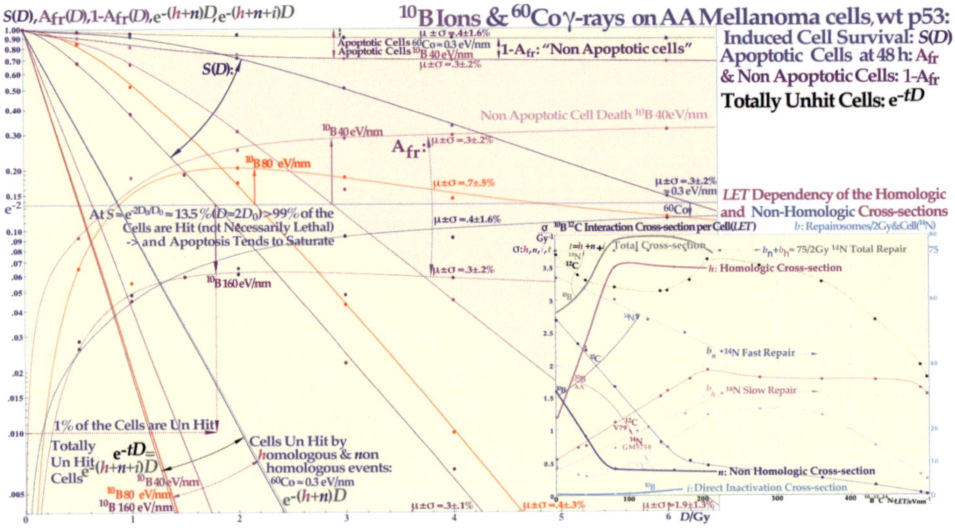

Figure 4.

*The cell survival, the cell fractions that are totally un-hit by the beams, and the apoptotic and non-apoptotic death over the LET range 0.3-40-80-160 eV/nm from ^{60}Co and boron ions. The cell survival shows a gradual increase in steepness with increasing LET whereas the A_{fr} has its maximum at a dose causing around 13.5% cell survival as indicated by the arrows. For the two lowest LET's, the non-apoptotic cells, upper dashed curves, and the clonogenic survival are practically tangential at low doses indicating apoptosis is the preferred way of cell death before p53 is phosphorylated at its Serine 15 and 20 sites at ≈½ Gy (**Figure 2**). The shaded area is due to non-apoptotic cell death for 40 eV/nm boron ions showing the domination of apoptotic cell death at the low LET's as the shading is lost at low doses. The survival data are also used in **Figure 14** on a linear scale to derive the secondary cancer induction probability. The associated LET variation of the non-homological and homological interaction cross-sections n and h for DNA repair after ^{10}B irradiation but also for ^{12}C ions, as shown by the insert [1, 4]. A comparison with the variation of the fast b_n and slow b_h repair of repair some foci after ^{14}N irradiation (dashed lines, right scale, data derived from [4, 17]). The homological cross-section h increases very fast with the LET for ^{10}B ions due to rapidly narrowing δ-electron cores and so is the associated reduction of the n cross-sections.*

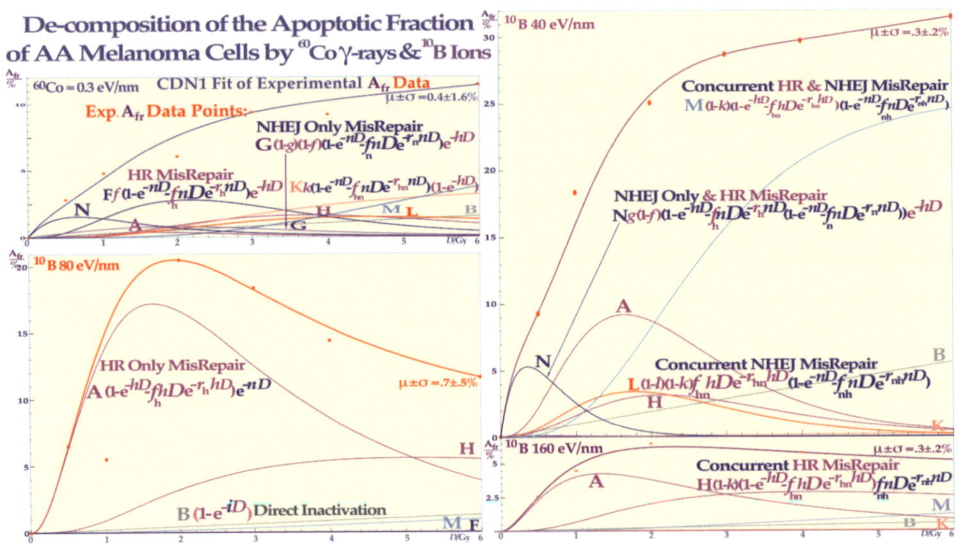

Figure 5.

The LET variation of the apoptotic fraction contributions of the eight key misrepair processes A-N indicating a relative apoptotic effectiveness (RAE) of about 3.4 for low LET boron ions around 40 eV/nm, whereas the peak relative biologic effectiveness (RBE) is about 3.5 but closer to an LET of about 140 eV/nm (Modified from [1]).

with additional BeO forming toxicity in the tumor but also in normal tissues cf. **Figures 11** and **13**). Everywhere else they mainly induce low-ionization density and LET-type effective DNA repair processes as seen in the right part of **Figure 2**. The surprisingly popular proton therapy has an almost negligible such augmented local high biological tumor effect, helium has some, lithium, beryllium, and boron have more and more, and finally the already rather promising carbon ions have a little too much, both in the entrance region with mainly normal tissues and the fragmentation tail beyond the tumor, and not least up to as far as 5 cm in front of the Bragg peak (cf **Figure 12**). Boron ions are therefore most likely the optimal ion for medium to large tumors whereas for small to medium size tumors and not least pediatric tumors a combination with lithium ions may be the most optimal for clinical use (cf **Figures 2** and **15**)! Interestingly, the lightest existing positron emitter Boron 8, makes it possible to immediately visualize the delivered dose to the patient by Positron Emission Tomography (PET) and verify that optimal dose delivery is achieved in the tumor region.

2.5 Relative biological effectiveness

With increasing atomic number, the penumbra gets narrower and the longitudinal range straggling is also lower, so more of the energy is deposited in the tumor by light ions of increasing atomic weight and Relative Biological Effectiveness (*RBE*) as seen in **Figure 6**. The high energy deposition density at the end of the ion range (the Bragg peak) is caused by a velocity resonance increasing the energy transfer to the tumor cells when the speed of the ion is close to the speeds of the orbital electrons of the tumor tissue and there is an increased probability for high energy transferred from the multiply charged ions to the electrons as they travel longer distances together toward the end of the particle range. The resulting peak is seen in **Figures 2, 6–8, 11** and **12**. When the atomic weight gets too high, the amount of particle fragments increases, so the dose beyond the Bragg peak gets high too, as seen in **Figures 12** and **13**. Ions heavier than carbon should therefore be used very carefully with sensitive normal tissues in front of and beyond the tumor.

3. Physical, biological, and clinical properties of light ion beams

Compared to low *LET* photon, electron, and proton beams, the clinical properties of light ion beams for radiation therapy are much more versatile and complex as discussed in more detail below. A large part of the detailed specific information is presented in graphical form in the Figures and their captions for simplicity and clarity.

3.1 Ion pencil beams

The physical and biological properties of narrow beams of the six lightest ions when penetrating water to a depth of 30 cm by their central axis energy deposition density profiles already shown in the lower row of **Figure 1** in 2D are described in 3D. The influence of multiple Columb scattering and longitudinal range straggling on the Bragg peak of the pencil beams is clearly shown in **Figure 7** (see also **Figures 8–10** below). With protons, the dose to normal tissues in front of the tumor is twice the

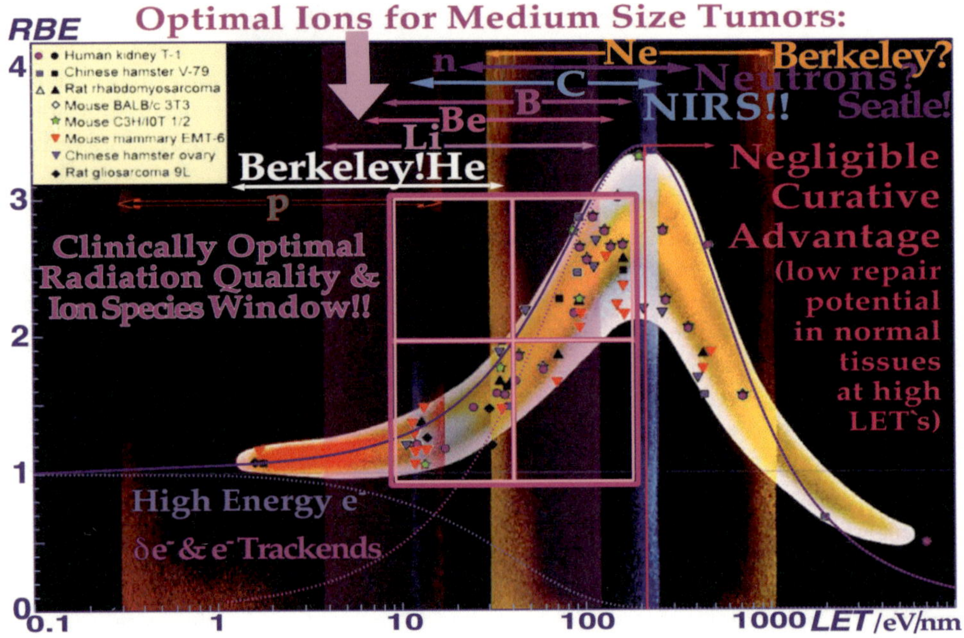

Figure 6.

Comparison of the RBE and LET ranges available with protons, lithium, boron, carbon, and neon ions is shown. It is seen that the range from lithium to carbon ions is most interesting, especially for hypoxic tumors. In general, LETs beyond the RBE peak should be avoided to minimize normal tissue damage in the entrance and plateau region (C-Ne, cf. **Figure 2***). As the cross-section saturates (cf insert in* **Figure 3***), the relative biological effectiveness RBE reaches a maximum since the cross-section cannot increase with the LET anymore, and at higher LETs the biological effectiveness decreases because of an increased probability of radical—radical recombination as secondary electrons are generated more and more closely together. Furthermore, the "overkill" effect implies that multiple kill events are equal to a single kill (you can only die once). The dashed and solid curves [18] describe the average response of the multiple experimental data sets very well [19].*

tumor dose due to significant multiple scatter whereas it is only a small fraction of the tumor dose for the light ions of lithium, boron, and carbon. This makes the local increase in dose to a small region of normal tissues in front of the tumor about five times larger than the increase in tumor dose that may be needed for tumor cure. This demonstrating that for each dose addition to a small part of a tumor volume will require that about five times more doses have to be given to normal tissues in front of the tumor with protons as compared to other light ions from lithium to boron. This is a severe dose delivery disadvantage for small radiation-resistant tumors and extra beam portals may be needed to avoid normal tissue damage with protons. It may be less of a dosimetric problem for large tumors where the broad beam Bragg peak dose level is better established. However, large tumors often have extensive hypoxic regions so protons are not generally the radiation modality of choice, and lithium to boron and even carbon ions are more often indicated for larger hypoxic tumors. The shapes of the so-called narrow pencil beams in **Figure 7**, therefore, are a kind of figure of merit when using inverse biologically optimized treatment planning trying to maximize the patient fraction that is cured without severe damage to normal tissues [3]. A high distal the so-called Bragg peak is therefore important and so is a low dose in the entrance and exit regions to minimize normal tissue damage in front of and not least behind the tumor by nuclear fragmentation as seen in even more detail for Boron and Carbon in **Figure 12**.

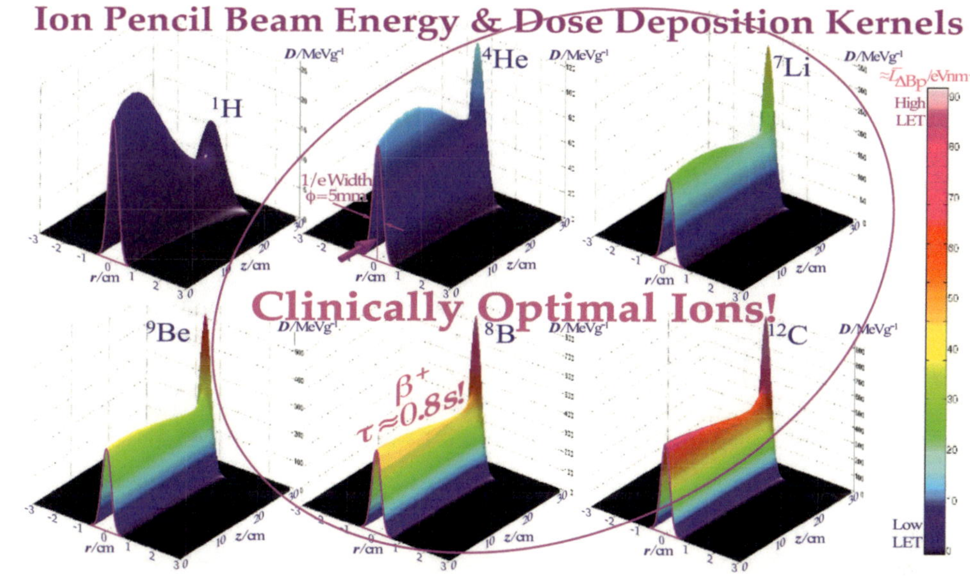

Figure 7.
3D Illustration of the clinical value of different 5 mm 1/e width light ion pencil beams for biologically optimized therapy planning. For carbon ions and heavier, the increasing LET in the plateau-type entrance region has to be considered more carefully when maximizing the probability of reaching a complication-free cure. The color scale illustrates the ionization density (LET) and consequently the increased biological effect in the tumor, which comes as a very important biological advantage on top of the physical dose distributional advantage shown in the figure.

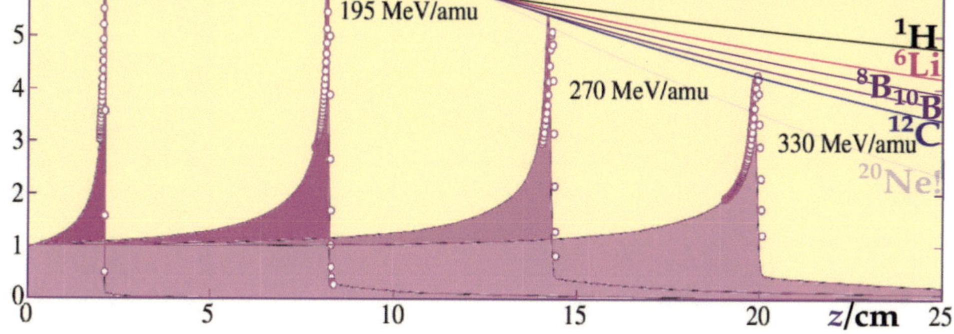

Figure 8.
As the cross-section increases with nuclear size light ion attenuation will be less in the normal tissues of the patient, to more effectively reach a deep sited tumor! Hydrogen would thus be optimal from this point of view but the biological effect is minimal as seen in **Figures 2** *and* **4***, making He, Li, and B more interesting!.*

3.2 Exponential ion attenuation

Like Photons, Light ions are attenuated exponentially but by nuclear reactions rather than by photoelectric, Compton, pair, and photonuclear reactions. Protons are only weekly attenuated, less than 50 MV photons, whereas carbon ions are similarly attenuated as 16 MV photons and neon more like 4MV and almost ^{60}Co. Part of the clinical problem with Neon ions is very clear as they are severely attenuated at large depths. **Figure 6** shows the other part of their problem as their entrance regions have a very high RBE and they reach far into the overkill region at depth as also seen in **Figure 11**. The merits of the light ions from lithium to boron are clearly shown in the figure all with an attenuation less than that of a 45 MV Betatron beam. Furthermore, to reach 25 cm of tumor depth a 400 MeV/u carbon ion beam may be needed whereas 300 MeV/u boron may suffice, thus requiring a smaller and less costly cyclotron for beam production as seen in the figure.

3.3 Particle multiple columb scattering

The increase in the penumbra (distance between 80 and 20% isodose lines) as a function of depth in water is shown in **Figures 9** and **10**. The Multiple Columb Scattering in the patient will increase the penumbra substantially with depth since the scattering power increase with decreasing energy. On top of these, multiple scatter contributions ($r_{1/e} = \sigma_r$) the part from the initial effective source size (σ_0) of the intrinsic accelerator beam should be added in quadrature as they stem from statistically independent processes so the total standard deviation is given

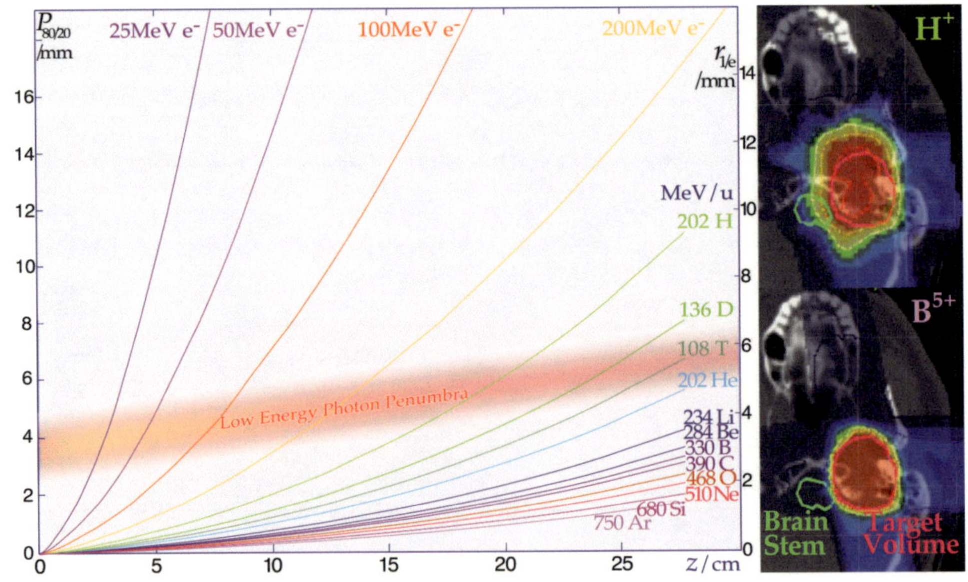

Figure 9.
With electrons and protons, the penumbra width is better than for photons at shallow depths (<5 cm), whereas light ions from helium and beyond are needed to get significant improvements compared to photons at large tumor depths. The insert shows the clinical advantage of a sharp penumbra in the neighborhood of organs at risk. The brainstem in this case is almost totally avoided with carbon or boron ions but not with protons (courtesy Jürgen Debus, Heidelberg).

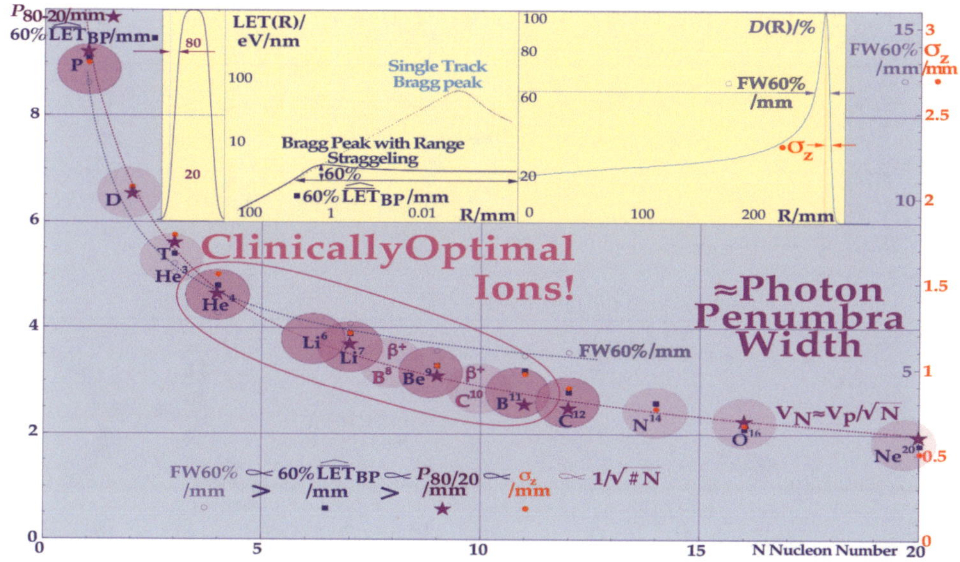

Figure 10.
The variation of the lateral penumbra and longitudinal Bragg peak width and range straggling for light ions of increasing nucleon number. A sharp reduction to 50% of the wide half-value width of protons is seen for helium and about one-third for lithium and about a quarter for boron and beyond.

by $\sigma_{tot}^2 = \sigma_r^2 + \sigma_0^2$. This initial part is included in **Figure 7**, where σ_0 = 2,5 mm (1/*e* width = 5 mm). The clear improvement in tumor coverage and normal tissue avoidance is seen in **Figures 9** and **10**, where also the reduction in penumbra width is seen beyond helium ions. In **Figure 10** the initial very steep reduction in the lateral penumbra and the longitudinal range straggling when going from protons to helium and more gradual going from helium to lithium, boron and carbon is clearly demonstrated. Since the penumbra of helium ions (α-particles) is already half of that for protons, helium ions are really the particle of choice in the low *LET* region as they in addition have very clear biological advantages not least in hypoxic tumors (cf **Figures 2, 6, 7, 9** and **11**).

3.4 Lateral penumbra width and longitudinal range straggling

The lateral multiple scattering and the penumbra decrease in almost the same fashion as the longitudinal straggling measured at the 60% width of the high *LET* region of the Bragg peak and they are closely proportional to the inverse square root of the nucleon number as seen **Figure 10**. The full 60% Bragg peak width is approximately the half width of these former quantities as seen in the right insert [20]. One may ask why there is such a large difference in biological effect between protons and other light ions (see **Figure 7**) even though their normalized broad beam dose distributions are fairly similar (cf **Figures 8** and **11**). This is mainly due to the same phenomenon that reduces their penumbra width and longitudinal range straggling with increasing nucleon number as seen in the middle straggling insert in **Figure 10**, reducing the single track LET from about 70 eV/nm over some 10 μm to a mean value of about 5 eV/nm in the few mm of range straggling at the Bragg peaks in clinical beams. This also shows up in the small *RBE* variation of protons in **Figure 11**.

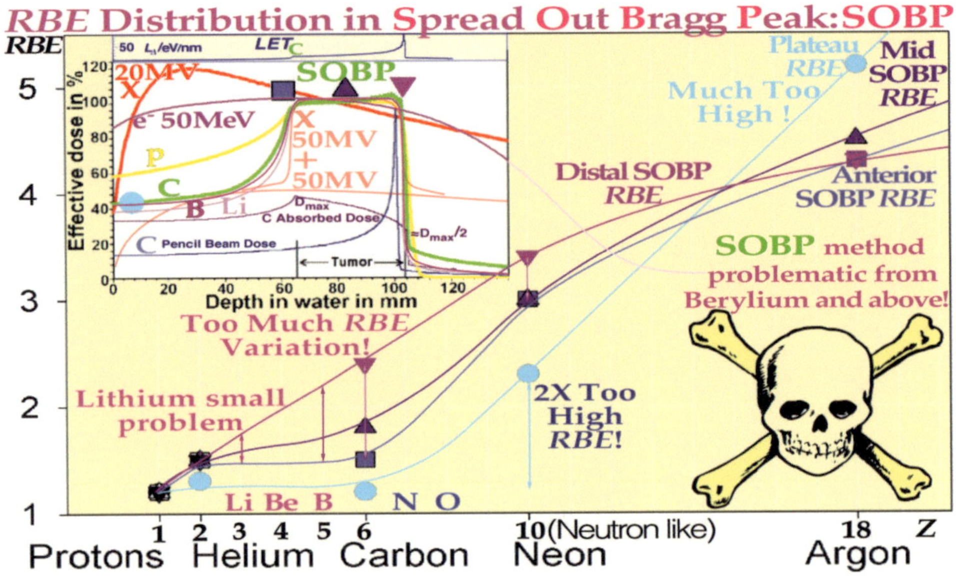

Figure 11.

The Variation of the biological effect over the SOBP will make the dose at the distal target volume low and LET high making the risk for microscopic cold spots high and increasing the risk for a recurrent tumor [5]. With high energy electrons and photons two perpendicular beams make a better high-dose dose distribution than a proton SOBP even if the low dose is lower, making Li, B, and C ions most interesting from a therapeutic point of view. With B and C ions, the method with two different intensity-modulated beams will eliminate the single beam SOBP problem as shown in **Figure 15**.

3.5 The *RBE* variation in the beam

The traditional way to make a uniform dose in an extended tumor volume is to use the so-called Spread Out Bragg Peak (SOBP) method first developed for protons at Berkeley and Uppsala where the energy and thus the range was modulated to get a uniform dose in the tumor volume [21]. This works well for protons with almost negligible biological effect variation with depth as seen in **Figure 11**. However, with Boron and Carbon ions there will be a substantial RBE variation with depth so the distal dose will be about half of that at the anterior part of the tumor generating an approximately uniform mean cell kill but a about twofold variation in biological effect as seen in **Figure 11**. This is far from desirable for a uniform tumor or even a heterogeneous tumor as seen from the large RBE variation measured for carbon ions in **Figure 11**. Combining lithium and boron ions in suitable ratios is better (**Figure 15**).

3.6 LET variation in beams

To clearly show the physical and biological differences between boron and carbon ion beams their partial ionization density contributions along with their absorbed dose distributions are shown in **Figure 12**. Even if their physical dose distributions are quite similar as already seen in **Figure 7**, their local ionization densities and LET's are rather different with the whole entrance region of carbon being of medium *LET*. With boron, this region is mainly low *LET* and the high *LET* region only extends ≈2 cm in front of the Bragg peak whereas it is about 5 cm for carbon ions. Furthermore, there is negligible elevated *LET* behind the Bragg peak of boron, so normal tissues both in front of and behind the tumor are much less damaged by boron therapy which is an

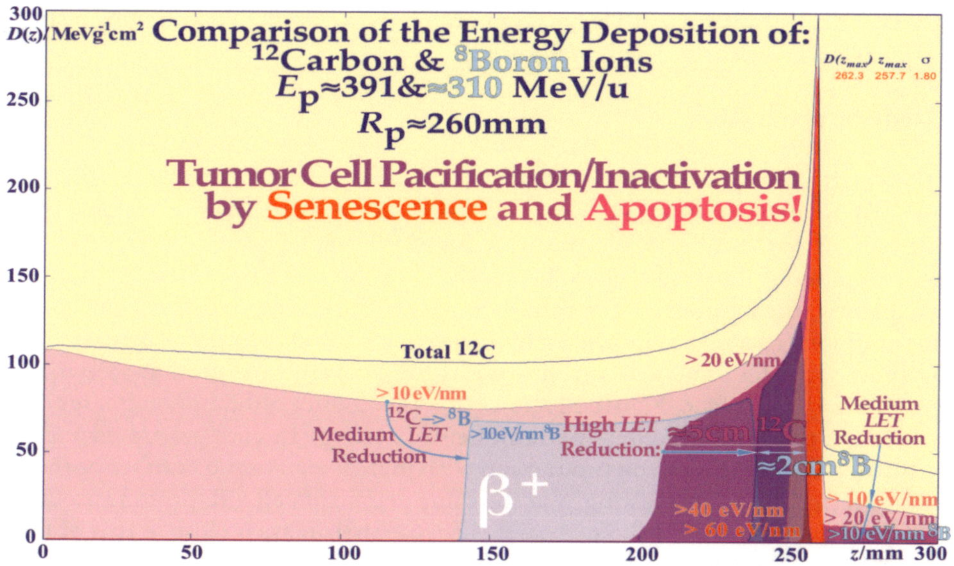

Figure 12.
Carbon and boron ion beam depth distribution of the energy deposition density. It is clearly seen that the adverse Biological effect in the entrance and fragmentation tail regions are significantly reduced by boron ions. With sensitive organs at risk in front of and behind the tumor volume, the high LET reduction will also be a further important advantage of boron ions as they will generate more apoptosis in the tumor than carbon ions as seen in **Figure 16**.

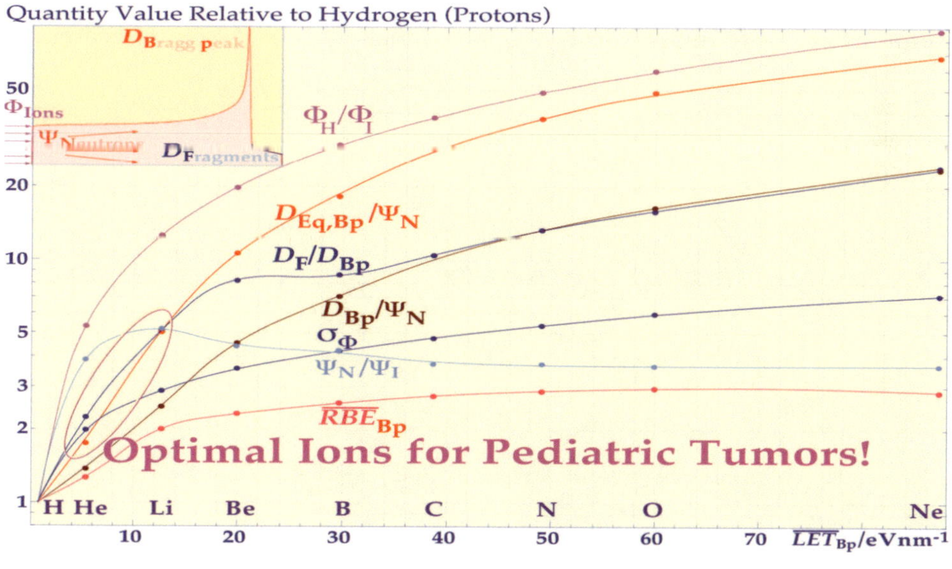

Figure 13.
Ion absorbed dose per neutron energy fluence. The increase of the RBE, the Bragg peak and fragmentation tail doses, D_{Bp} and D_F per unit neutron energy fluence, Ψ_N, and ion fluence, Φ_I, at tumor lethal doses of light ions as a function of the mean Bragg peak LET values are shown. All the values are given relative to those for protons and the curves clearly show how much higher the fluence of protons is compared to other light ions up to neon at a given biological effect. Furthermore, the increase in therapeutically effective dose in the Bragg peak per generated unit of neutron energy fluence, the ratios of tail fragment dose compared to the Bragg peak dose, the Bragg peak absorbed dose per neutron energy fluence, the increasing variance of the particle fluence as caused by the reduced fluence of ions at high LET s and RBEs, and the mean ion Bragg peak RBE are all shown in order from top to bottom.

important clinical advantage! This makes boron ions most suitable for mixed beam therapy as is more clearly shown in **Figure 15**.

3.7 Neutron production

Like high-energy photon beams light ion beams are always associated with a significant neutron production as seen in **Figure 13**. Due to nuclear reactions, the cross-section and number of neutrons produced increases with the increasing atomic number of the projectile as the number of fragmentation reactions and consequently, the number of neutrons increases with nuclear size. However, the number of ions needed to eradicate a hypoxic tumor simultaneously decreases rapidly since the *LET* and *RBE* increase with atomic charge and mass as seen in **Figures 6** and **13**. In fact, **Figure 13** shows that the fluence reduction increases both the dose and the equivalent dose per unit neutron energy fluence generated and thus steadily decreases the neutron production with increasing atomic number of the ion at tumor lethal doses. Interestingly, at the same degree of cell kill, protons produce the highest neutron energy fluence partly because their mass is close to that of the neutron. From beryllium and above, the fragmentation is rather high so helium and lithium are probably the most ideal ion species for the treatment of pediatric malignancies. From the point of view of neutron and light fragment production, the ideal ion is generally somewhere between lithium and boron as seen in **Figure 13** but avoiding beryllium's high fragmentation and potential toxicity. Interestingly, even if the cross-section per ion to produce neutrons goes up slowly with atomic number, about 5, 15, and 40 times higher fluencies of protons are needed compared to helium, lithium, and carbon ions, making the total neutron production the highest by protons at equally curative doses of light ion therapy. Both the mean absorbed

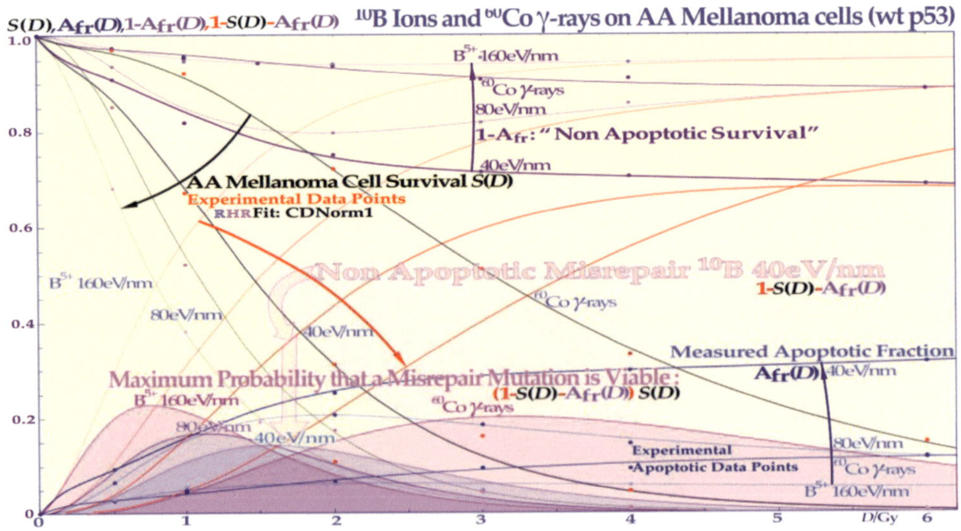

Figure 14.
The probability for secondary cancer induction as a function of the dose delivered to normal tissues. At low doses the risk of inducing a mutation is small, at high doses the probability to generate it is higher but so is the probability to eliminate it by the treatment. The risk is high in normal tissues between 0.5 and 6 Gy so this volume in the patient should be minimal! The LDA and LDHS of this TP53 intact cell line are clear! Interestingly, the risk is the smallest for the lowest LET boron ions due to HDA! CDN1: One-dimensional closest distance norm (not least square!). Based on data from [1, 5].

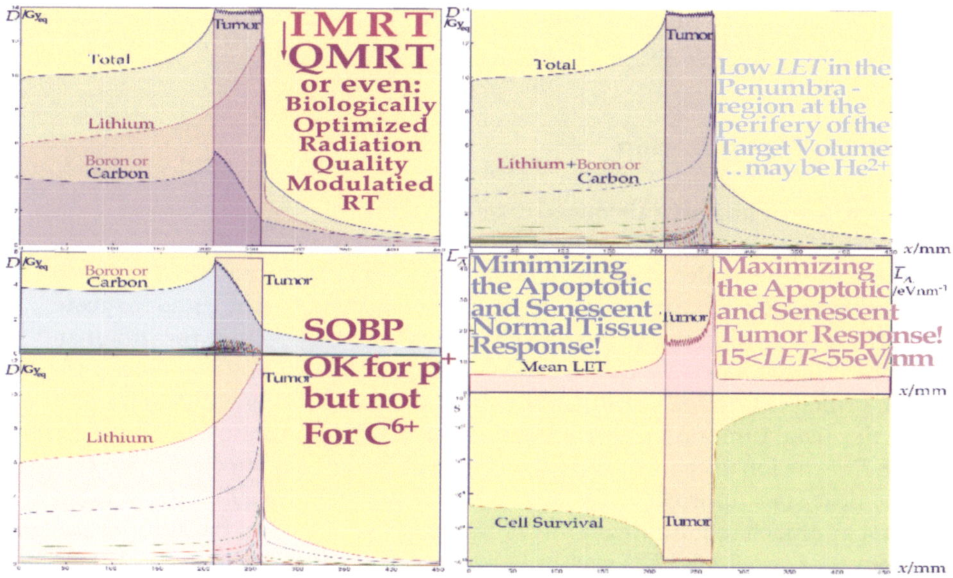

Figure 15.
A quasi-uniform absorbed dose and cell kill distribution is generated between about 21 and 26 cm of depth by combining lithium and boron or carbon ions in suitable ratios to make the cell kill and survival quasi-uniform. The small local fluctuations in absorbed dose are due to a somewhat too large longitudinal range modulation (~3 mm) used to clearly illustrate the applied mechanism combining lithium and boron or carbon ion Bragg peaks at each depth interval. The different panels show the total absorbed dose and the boron or carbon doses and the lithium dose in the upper row, whereas the cell survival and mean LET distribution is shown below. Interestingly, by combining lithium and boron or carbon, a uniform biological effect, survival, and absorbed dose can be obtained both for uniform tumors and optimal biological effect modulation for heterogeneous tumors [23].

dose (D_{Bp}) and dose equivalent ($D_{\mathrm{Eq,Bp}}$) at the Bragg peak per unit energy fluence of neutrons generated (Ψ_{N}) increase rapidly with the atomic number as seen in **Figure 13**, indicating that the absolute neutron production reduces steadily with atomic number. The dose of fragments is unfortunately also increasing. Helium and lithium ions are therefore indicated to be most optimal for pediatric tumors where the neutron fluence, dose and biological effect on normal tissues surrounding the tumor should really be minimized to reduce as far as possible the neutron-generated risk for secondary cancers.

3.8 Secondary cancer induction

To further illustrate the power of quantifying apoptosis, **Figure 14** estimates the probability of inducing secondary cancer based on experimental cell survival and apoptosis data [1: Figures 7+9]. It is unlikely that the Apoptotic fraction will contribute to secondary cancer induction (except possibly in TP53 mutant cell lines) so it is useful that this fraction can be estimated by the new RHR formula and be removed from other forms of misrepair to more accurately describe the cells that are potentially capable to generating secondary cancers. This cell fraction, as seen in **Figure 4**, has its peak in the 1-3 Gy range so in radiation therapy optimization, it is desirable to minimize this volume as much as possible in normal tissues. **Figure 14** show the maximal risk is smallest for low LET ions (blue shaded), the real risk may be in the order of 5% of the values in the Figure. This effect is a contraindication for very many beam portals in intensity modulated radiation therapy e.g., using “rapidarc”, “volumetric arc", and "tomotherapy" like methods on non-seniors, that may have time to develop secondary cancer after some 20 years cf [22].

4. Treatment optimization

4.1 Generation of uniform absorbed dose and cell kill

It is better from a microdosimetric point of view to generate a rather uniform microscopic energy deposition density on the cellular scale in the tumor to avoid dosimetric cold spots in tumor clonogens [5] that could repopulate the tumor and start with a slightly lower *LET* at the anterior tumor edge as shown in **Figure 15**. This could in principle be done in two different ways, either starting with a suitable Bragg peak at the distal tumor edge and gradually increasing the atomic weight and Bragg peak *LET* of the ion used or by just mixing two different ions species so that the mean *LET* and dose stays approximately constant as shown in **Figure 15**. Interestingly, this latter approach requires that most of the Bragg peak dose in the anterior part of the tumor is of very high *LET*, such as boron or carbon ions, whereas the distal part mainly requires a lower *LET* such as helium or lithium ions. In fact, mixing lithium and boron or carbon ions may be the optimal way to achieve close to ideal microscopic energy deposition density distribution for medium-to-large size tumors whereas several small oligo-metastasis are best treated by lithium ions alone. The survival was calculated for simultaneous irradiation whereas the mean *LET* is based both on lithium and boron or carbon ion components, but the *LET* variation was low, and

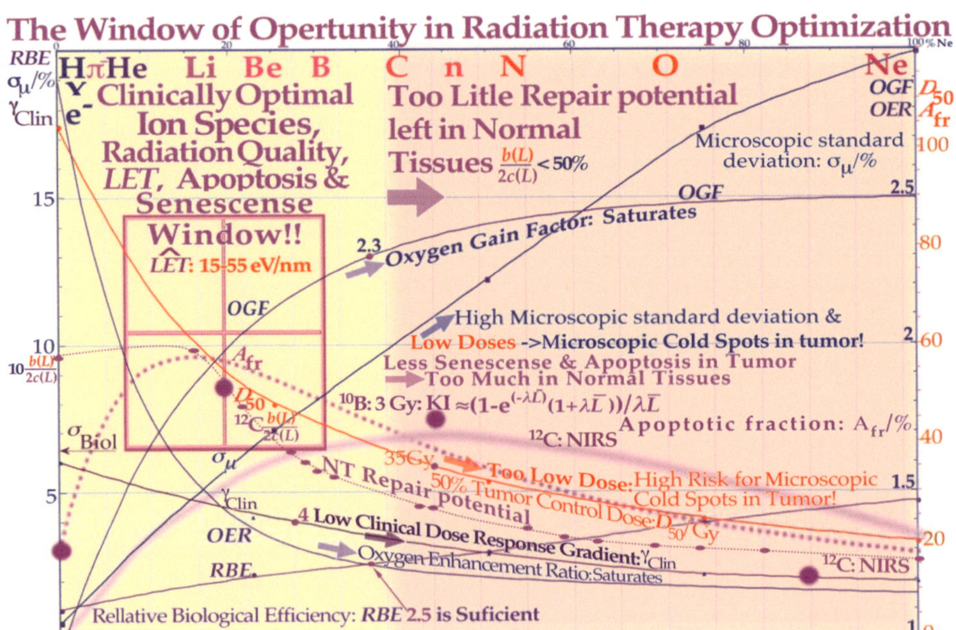

Figure 16.

The LET variation of key biological parameters that influence the clinical value of radiation beams showing that the optimal window of opportunity in radiation therapy optimization is located between about 15 and 55 eV/nm or He-B ions. The underlying data are collected from Berkeley, NIRS, and Karolinska.

it peaks just downstream from the distal part of the tumor where the dose is a little low. The figure also illustrates the principle of Quality Modulated Radiation Therapy (QMRT) where the absorbed dose and biological effect and radiation quality can be optimally modulated instead of just the Intensity (IMRT) e.g., over a severely hypoxic tumor [23, 24].

4.2 Selection of optimal particle and LET

Based on the cell survival curves in **Figure 4** and apoptosis curves in **Figure 5**, one may ask which ion and LET should be used to maximize the complication-free cure of the radiation treatment. The very flexible method of mixing two ions with largely different ionization densities as shown in **Figure 15**, such as He or Li and B or C, rather than using Spread-Out Bragg Peaks [21], which work well only for protons as seen in **Figure 11**. The SOBP method generates strong variations in ionization density and absorbed dose. The variation is a factor of approximately two over the whole target volume for carbon ions, with a too low LET at the anterior end and too low dose and too high LET at the distal part of the target volume as seen in **Figure 11**. This significantly increases the risk of microscopic cold spots at the distal part of the target volume [5]. The use of mixed-modality treatments can make the absorbed dose and biological effect almost constant (cf **Figure 15**), or even locally elevated, e.g., in regions of strong hypoxia [2, 23, 24]. The beam quality question will therefore be discussed in terms of the optimal particle species and mean LET, as shown in **Figure 16** for the LET range from 0.2 to 180 eV/nm covering X-rays-Ne ions as recently discussed [4]. "In this region the RBE increases steadily from 1 to approximately 4.5, the oxygen enhancement ratio (OER) decreases from approximately 3 to almost 1, the normal tissue repair potential $b/2c$ decreases from about 1 to ¼, the normalized clinical dose response gradient, γ_C, decreases from approximately 6 to 2 (cf **Figure 17**), the microscopic standard deviation in dose σ_μ increases from 1% to approximately 20%, the 50% tumor control dose, $D50$ decreases from almost 100 to 20 Gy for radiation-resistant tumors, the apoptotic fraction, Afr varies from almost 60 down to ≈3% and the oxygen gain factor (OGF) increases from 0 to 2.5. Importantly, above approximately 55 eV/nm, many of these factors become less advantageous for clinical use: the loss of sublethal DNA repair in normal tissues (see **Figure 4** with insert), the saturation of the OER and the OGF, and the reduced senescence and apoptosis in the tumor (**Figure 16**)". The increase in microscopic standard deviation (σ_μ) will decrease the clinical γ_C value and microscopic cold spots may appear as the standard deviation in dose delivery becomes more significant, as the therapeutic dose reaches ≈35 Gy and lower [5]. Consequentially, the senescence and apoptosis in the tumor decrease while it increases too much in normal tissues! Therefore, boron ions are more optimal than carbon ions (**Figure 12**), at least for medium-sized tumors, and lithium ions are the optimal particle for pediatric tumors, and their combination is ideal both for dose and radiation therapy biological effect optimization (**Figures 12** and **15**). It is very interesting to see that the apoptotic and senescent cell fractions have their maxima in the tumor at a much lower LET than the RBE, and it reaches approximately 60% at approximately 30 eV/nm for boron ions (dotted curve in **Figure 14**), whereas the RBE has its maximum at around 140 eV/nm. This is mainly due to the larger number of ions needed per unit dose compared to carbon ions (cf. **Figure 13**). Carbon ions reach only approximately 40% at approximately 80 eV/nm and they may produce unnecessarily high senescence, apoptosis, and LET in normal tissues [4] as seen in **Figure 6**. Optimal

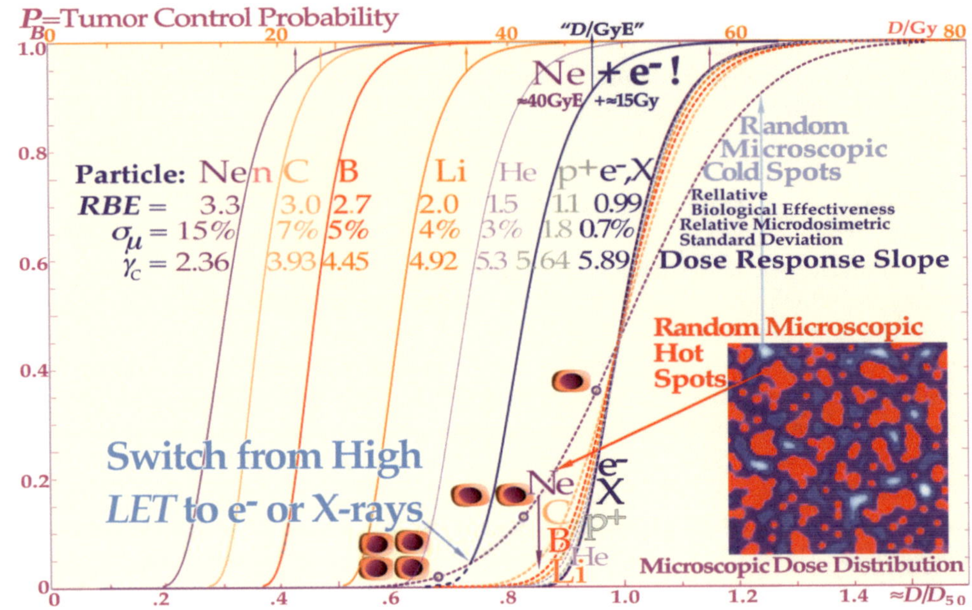

Figure 17.
Microdosimetry of tumor control. Description of the reduction in the normalized steepness, γ_C, of the shape of the tumor control curve for uniform cell line for the different microscopic standard deviation of radiation modalities as a function of the absorbed dose (upper scale) and approximately normalized to the 50% tumor control dose (≈Dose Eq, lower scale, dashed lines) to more clearly see the effect on the γ_C value as the microdosimetric relative standard deviation increases with the LET. Not only do the hot spots often in the form of dual double strand brakes (DDSB, cf. **Figures 2** *and* **3***) and cold regions get more extreme with increasing LET, but also the RBE is increasing, reducing the total dose about threefold with carbon, neutron, and neon, increasing the relative standard deviation and reducing the γ_C value more than one would like.*

cancer cell inactivation requires a treatment modality that preferentially induces senescence probably the most cost-efficient treatment to stop further cell cycling and block tumor growth. Actually, this is probably the mildest but still efficient endpoint to cure cancer [5, 10, 11, 25, 26], as it can make the clonogenic tumor cells lose their uncontrolled cell cycling ability as seen in **Figure 18**. The induction of autophagy (self-digestion) and apoptosis (programmed cell death, see **Figure 2**) may follow more severe DNA damage to minimize the cancer induction with more complex DNA damage and increased risk for severe mutations, e.g., with turned-on oncogenes or lost suppressor genes. Apoptosis and, even more so, senescence is probably the most optimal way to inactivate a tumor with minimal inflammatory response and without massive immediate apoptotic tumor decomposition [5, 25, 26]. Lithium ions, and its neighbors helium and boron, have an important and unique potential to induce apoptosis locally, mainly in a few mm-size volumes around their Bragg peaks in the tumor, and everywhere else, induce low-*LET* DNA repair facilitating molecular radiation therapy (see **Figure 2**). In repair terms, this means that NHEJ will dominate in the entrance and fragmentation tail of the ion, whereas HR will mainly try to handle the partly irreparable damage at the Bragg peak, which therefore should solely be placed into the tumor volume [1]. Thus, lithium ions are probably the optimal ion, at least for smaller tumors (see **Figures 2** and **11**, [10]) and pediatric patients. The major advantage of lithium ions is the low-ionization density in all normal tissues, largely inducing fast DNA repair while often inducing apoptosis and senescence only in the tumor volume. For the same reasons, medium size

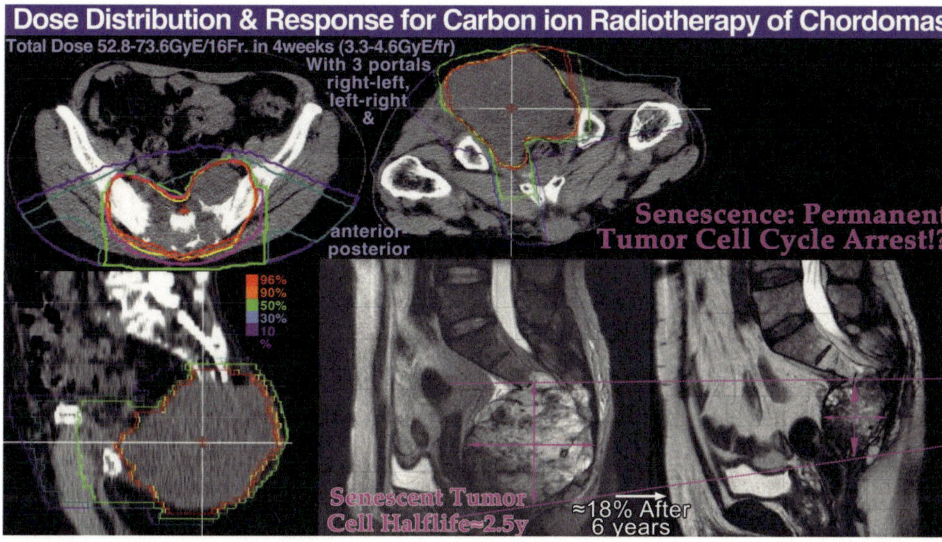

Figure 18.
Illustration of the dose delivery for a large pelvic chordoma as shown in the upper and lower left treatment plans. The lower right plain MR images show the gradual disappearance of the chordoma 6 years after the treatment, probably due to a totally senescent response without traces of tumor growth. This treatment was made with carbon ions but could most likely be done with an advantage with boron ions as seen by the B and C ion LET data and apoptosis in **Figures 7–17**! *(Courtesy: Hirohiko Tsujii, NIRS, Chiba, Japan).*

tumors are probably best treated using boron ions, but this may sometimes require an extra beam portal compared to carbon ions, especially for larger hypoxic tumors. Interestingly, for radiation resistant commonly TP53 mutant tumors (cf **Figure 2** lower left insert) the p53 reactivating compound APR-246 can increase the apoptotic-senescent response and increase the effect of Reactive Oxygen Species (1: Figures 6+17, [5, 25, 27]) and will work well adjuvant also with Boron-ions. Furthermore, ^{8}B is our lightest and best direct PET emitter for concomitant online dose delivery imaging, although perhaps a little complex to make [28], and with a very short half-life of ≈0.8 sec so the PET camera really needs to be in the treatment room and used between ion accelerator pulses while treating the patient with the advantage of a rapid malfunction response time (10, 29, Figure 8.37, 8.38f).

4.3 Influence microdosimetric heterogeneity on tumor control

The shape of the Tumor Control curve is shown in **Figure 16** for a uniform cell line with different radiation modalities as a function of the absorbed dose (upper scale) and approximately normalized to the 50% tumor control dose (≈Dose Equivalent, lower scale) to more clearly see the effect of increasing microscopic heterogeneity as measured by the microdosimetric relative standard deviation (σ_{μ}) with increasing ion mass, *LET* and *RBE*. Not only do the hot spots often in the form of Dual Double Strand Brakes (DDSB's, cf. **Figure 18**, [4: **Figure 2**]) and cold regions get more extreme with increasing *LET*, but also the *RBE* is increasing so the reduction of the total dose is about threefold with carbon, neutron, and neon increasing the relative standard deviation and reducing the clinical dose-response slope γ_C severely. A steeper tumor control curve generally increases the therapeutic window of radiation therapy since the absorbed dose distribution and the associated therapeutic effect over the therapeutic window can be modulated with greater efficiency with a steeper tumor cure and normal tissue damage curves [5]. Also with boron ions the last ≈10 GyE should therefore be delivered by low LET beams [5].

4.4 Apoptosis and senescence

Since apoptosis is nature's way to eliminate unwanted cells during the development of practically all organs and is therefore not generally associated with any inflammatory responses accompanying the more common necrotic type of cell kill. In addition, permanent cell cycle arrest or senescence and apoptosis are increasingly induced by light ions. Apoptosis through caspase 3 may cause accelerated tumor cell repopulation after a non-curative treatment [9]. Senescence may therefore be the most desirable endpoint of cancer therapy as the tumor cells lose their reproductive ability and are then slowly disappearing depending on the remaining cellular lifetime or half-life ≈2.5 years. As seen in **Figure 18** the tumor was reduced to about half the diameter or 10–15% of the initial volume 6 years after the treatment. Due to the fact that more ions per unit dose and cell kill are needed at medium to low ionization density, the more effective apoptotic and senescent response is obtained at an ionization density of around 20 eV/nm to 40 eV/nm as shown theoretically and experimentally in **Figures 15** and **16** [29]. Interestingly, helium, lithium and boron combine a high local apoptotic and senescent tumor cell inactivation only a few mm around their Bragg peaks and can thus be regarded as the ultimate stereotactic and conformal radiation modality (cf **Figures 1, 7, 12** [29–33]).

5. Conclusions

With boron ions the clinical experience from NIRS (National Institute for Radiological Sciences, Chiba) and other centers in Japan and Germany with carbon is an invaluable asset. A very promising development seen at NIRS is the very effective sterilization of large tumor masses (10–15 cm) such as pelvic chordomas as shown in **Figure 18** [29, 31–35]. These tumors show a gradual disappearance down to 10% and below the initial volume about 5 years after the treatment, probably due to a persistent senescent response likely to be even more effective with boron ions [35], and without evidence of any further tumor growth. The clinical value of light ion beams is discussed in further detail in a number of recent references [2, 5, 10, 19, 20, 29–34, 36]. When the above methods are brought into clinical use a mean tumor cure as high as 80% should be possible, and even more if the new therapeutic dose delivery principles [5], advanced dose fractionation [4: Figure 21, 5], and early tumor detection [10] and malignancy determination methods [37] come into more regular use.

Author details

Anders Brahme
Department of Oncology-Pathology, Karolinska Institutet, Stockholm, Sweden

*Address all correspondence to: andersbrah@gmail.com

References

[1] Brahme A. Quantifying cellular repair, misrepair and apoptosis induced by boron ions, gamma rays and PRIMA-1 using the RHR formulation. Radiation Research. 2022;**198**:271-296

[2] Brahme A, editor. A Brief Introduction to The Development of Radiation Therapy Optimization. Ch 1 In: Biologically optimized radiation therapy. Singapore: World Scientific Publishing; 2014:1-15

[3] Brahme A, Löf J. Fundamentals of physically and biologically based radiation therapy optimization In: Brahme A. editor. Comprehensive BioMedical Physics. Vol. 9 Ch 12. Elsevier Oxford: Major Reference Work; 2014. p. 271-313

[4] Brahme A. A DNA repair-based model of cell survival with important clinical consequences. Radiation Research. 2020;**194**:202-235

[5] Brahme A. On the Ultimate Biological Optimization Steps of Curative Radiation Oncology. Cancers, In Press; 2023

[6] Buscemi G, Perego P, Carenini N, Nakanishi M, Chessa L, Chen J, et al. Activation of ATM and Chk2 kinases in relation to the amount of DNA strand breaks. Oncogene. 2004;**23**:7691-7700

[7] Enns L, Bogen KT, Wizniak J, Murtha AD, Weinfeld M. Low-dose radiation hypersensitivity is associated with p53-dependent apoptosis. Molecular Cancer Research. 2004;**2**:557-566

[8] Short SC, Bourne S, Martindale C, Woodcock M, Jackson SP. DNA damage responses at low radiation doses. Radiation Research. 2005;**164**:292-302

[9] Zhao M, Wang Y, Zhao Y, He S, Zhao R, Song Y, et al. Caspase-3 knockout attenuates radiation-induced tumor repopulation via impairing the ATM/p53/Cox-2/PGE2 pathway in non-small cell lung cancer. Aging. 2020;**12**:21758-21776

[10] Brahme A. High resolution molecular radiation therapy and tumor imaging for the 21st century. Journal of Nuclear Medicine Radiology & Radiation Therapy. 2016;7:1-311. DOI: 10.4172/2155-9619.1000311+

[11] Brahme A. Development of Highly Specific Molecular Cancer Therapy with the Lightest Ions. Book of Abstract Japan: 5th Takahashi Memorial International Symposium; 2007. p. 57

[12] Nakamura Y. Isolation of p53-target genes and their functional analysis. Cancer Science. 2004;**95**:7-11

[13] Williams AB, Schumacher B. p53 in the DNA-damage-repair process. Cold Spring Harbor Perspectives in Medicine. 2016;**6**:a026070

[14] Brahme A, Rydberg B, Blomqvist P. Dual spatially correlated nucleosomal double strand breaks in cell inactivation. In: Goodhead DT, O'Neill P, Menzel HG, editors. Microdosimetry: An Interdisciplinary Approach. Cambridge, UK: The Royal Society of Chemistry; 1997. pp. 125-128

[15] Frankenberg D, Frankenberg-Schwager M, Harbich R. The contribution of OH* in densely ionizing electron track ends or particle tracks to the induction of DNA double strand breaks. Radiation Protection Dosimetry. 1990;**31**:249-252

[16] Stenerlöw B, Höglund E, Carlsson J, Blomquist E. Rejoining of

DNA fragments produced by radiations of different linear energy transfer. International Journal of Radiation Biology. 2000;**76**:549-557

[17] Ward JF. DNA damage produced by ionizing radiation in mammalian cells: Identities, mechanisms and reparability. Progress in Nucleic Acid Research and Molecular Biology. 1988;**35**:95-125

[18] Brahme A. Accurate description of the cell survival and biological effect at low and high doses and LETs. Journal of Radiation Research. 2011;**52**:389-407

[19] Blakely EA, Chang PY. Biology of charged particles. Cancer Journal. 2009;**15**:271-284

[20] Kempe J, Gudowska I, Brahme A. Depth absorbed dose and *LET* distributions of therapeutic 1H, 4He, 7Li and 12C beams. Medical Physics. 2007;**34**:183-192

[21] Raju MR. Heavy Particle Radiotherapy. London: Academic Press; 1980

[22] Hall EJ, Wuu CS. Radiation-induced second cancers: The impact of 3D-CRT and IMRT. International Journal of Radiation Oncology, Biology, Physics. 2003;**56**:83-88

[23] Brahme A, Ågren AK. On the optimal dose distribution for eradication of heterogeneous tumors. Acta Oncologica. 1987;**26**:377-385

[24] Brahme A, Källman P, Lind BK. Optimization of proton and heavy ion therapy using an adaptive inversion algorithm. Radiotherapy and Oncology. 1989;**15**:189-197

[25] Nardella C, Clohessy JG, Alimonti A, Pandolfi PP. Pro-senescence therapy for cancer treatment. Nature Reviews. Cancer. 2011;**11**:503-511

[26] Gorgoulis V, Adams PD, Alimonti A, Bennett DC, Bishop O, Bishop C, et al. Cellular senescence: Defining a path forward. Cell. 2019;**179**:813-827

[27] Bykov VJN, Eriksson SE, Bianchi J, Wiman KG. Targeting mutant p53 for efficient cancer therapy. Nature Reviews. Cancer. 2018;**18**:89-102

[28] Rubbia C, Ferrari A, Kadi Y, Vlachoudis V. Beam cooling with ionization losses. Nuclear Instruments and Methods in Physics Research A. 2006;**568**:475-487

[29] Vreede P, Brahme A. Development of biologically optimized radiation therapy: Maximizing the apoptotic cell kill. Radio Science. 2009;**52**(7):31-52. Available from: http://www.nirs.go.jp/info/report/rs-sci/pdf/200907.pdf

[30] Brahme A, Svensson H. Physical, biological and clinical background for the development of biologically optimized light ion therapy. In: Brahme A, editor. Biologically Optimized Radiation Therapy. World Scientific Publishing: Singapore; 2014. pp. 499-648

[31] Brahme A. Potential developments of light ion therapy The ultimate conformal treatment modality. Radio Science. 2009;**52**(2):8-31. Available from: http://www.nirs.go.jp/info/report/rs-sci/pdf/200902.pdf

[32] Brahme A. Optimal use of light ions for radiation therapy. Radio Science. 2010;**53**(8-9):35-61. Available from: http://www.nirs.go.jp/publication/rs-sci/pdf/201008.pdf

[33] Imai R, Kamada T, Sugahara S, Tsuji H, Tsujii H. Carbon ion radiotherapy for sacral chordoma. The British Journal of Radiology. 2011;**84**:S48-S54

[34] Imai R, Kamada T, Araki N. Carbon ion radiation therapy for unresectable sacral chordoma: An analysis of 188 cases. International Journal of Radiation Oncology, Biology, Physics. 2016;**95**:322-327

[35] Kumar S, Suman S, Fornace AJ Jr, Datta K. Space radiation triggers persistent stressresponse, increases senescent signaling, and decreases cell migration in mouse intestine. Proceedings of the National Academy of Sciences of the United States of America. 2018;**115**:201807522

[36] Yamada S, Yamamoto N, Koto M, Imai R, Kasuya G, Okonogi N, et al. Current status and perspective of carbon ion radiotherapy. Radioisotopes. 2019;**68**:395-402

[37] Rye IH, Lundin P, Måner S, Fjelldal R, Naume B, Wigler M, et al. Quantitative multigene FISH on breast carcinomas. Gen Chromo Ca. 2015;**54**:235-248

Chapter 2

Boron Removal by Donnan Dialysis According Doehlert Experimental Design

Ikhlass Marzouk Trifi, Lasâad Dammak, Lassaad Baklouti and Béchir Hamrouni

Abstract

Donnan dialysis is one of the membrane processes. It is based on the cross-exchange of ions having the same electric charge through an ion-exchange membrane. The removal of boron by Donnan dialysis was studied in this work. First, a preliminary study was conducted to determine the experimental field of operating parameters using two membranes (AFN and ACS). Then, a full factorial design was applied to investigate the influence of the operating parameters and their interactions on the boron removal. Response surface methodology using Doehlert design was adopted to predict the optimal conditions. This approach via experimental designs is more efficient than the conventional optimization approach (the "one-at-a-time" method) which is time-consuming and requires a large number of experiments.

Keywords: boron removal, Donnan dialysis, response surface methodology, experimental design, anionic exchange membranes

1. Introduction

Sources of boron in the environment are mainly natural or even anthropogenic. It can be found mainly in the form of boric acid or borate salts. The presence of boron compounds in water increases in a continuous and parallel way to industrial development. In fact, boron is used in a wide range of industrial applications such as glass and ceramic industry to produce borosilicate glass, insulation, fiberglass, and flame-retardant fiberglass [1]. Due to its high concentration of boron, natural water is often unsuitable for human consumption or agricultural use. Therefore, the harmful effects of boron on living organisms also increase, especially on plants, since boron manifests an important micronutrient-toxic boron duality [1]. Moreover, boron is a unique micronutrient in which overdose and underdose of boron supply cause toxicity and deficiency symptoms in plants, respectively. The level of boron in irrigation water exceeding 1 mg/L can affect the yield of sensitive crops (e.g., avocado and citrus fruits) [1]. Irrigation water with a very low boron content is required for certain metabolic activities, but when boron concentrations are increased to 4 mg/L [1], plants become poisoned, manifesting as yellow spots on leaves and fruits, and their decomposition speeds up and they die [2]. According to the World Health Organization (WHO), drinking water should have a boron content of less than 0.3 mg/L because humans may also be poisoned by excessive boron levels [3, 4]. Consequently, several studies have focused on boron removal such as adsorption [5, 6], electrocoagulation [7, 8], or membrane processes namely electrodialysis [9, 10], reverse osmosis [11, 12], nanofiltration [13, 14], microfiltration [15], ion-exchange [15, 16], membrane distillation [17], and Donnan dialysis which was the subject of this study [18, 19]. This membrane process referring to FG Donnan [20] uses ion-exchange membranes allowing cross-exchange of ions to separate or concentrate ionic species. Modern Donnan dialysis uses ion-exchange membranes to separate the two solutions that are involved in the transport process. With no external electric potential difference applied across the membrane, Donnan dialysis uses an ion-exchange membrane. Donnan dialysis uses the counterdiffusion of two or more ions through an ion-exchange membrane to achieve separation. Donnan dialysis is one of the easiest and most inexpensive membrane techniques. The chemical potential gradient between the constituents of two solutions separated by a membrane is what propels the process. When Donnan equilibrium is reached, the process of Donnan dialysis is complete. It involves the stoichiometric exchange of counterions, or ions with the same charge, over an ion-exchange membrane [21].

There is frequently no theoretical model or one that is quite sophisticated that links certain controllable variables (factors) to a response. In this situation, empirical data should be used to determine the link between the causes and the response. Box and Wilson first presented the Response Surface Methodology (RSM), a group of mathematical and statistical tools whose goal is to evaluate situations like the one given using an empirical model. The RSM was used to explore the removal of boron by Donnan dialysis. Herein, the removal of boron by Donnan dialysis was investigated using RSM approach. RSM is a useful method for process optimization when a number of independent factors and their interactions have an impact on the desired results. With RSM, multiple variables are tested concurrently with the least amount of trials possible in accordance with unique experimental designs built on factorial designs. This technology has the advantage over conventional approaches in that it requires less time and money. The goal of RSM is to characterize the behavior of a dataset in order to make statistical predictions. It is a set of statistical and mathematical tools enabling the adjustment of experimental data to a theoretical model expressed by a polynomial equation. The simultaneous variation of a large number of operating parameters at once, the reduced number of experiments, the detection of interactions between factors, and achieving the highest precision are all advantages of the RSM for optimization [22]. Recently, many statistical experimental design methods have been employed in optimization. Among them, Doehlert designs stand out, compared to other designs such as the central composite design or Box-Behnken design, by the reduced number of experiments and the possibility to assign a large or a small number of levels to the chosen variable. Indeed, this design presents variables with different numbers of levels. The variable with the stronger effect should be the variable with the highest levels if you want to learn as much as you can about the system.

Doehlert designs are also more efficient in mapping space: adjoining hexagons can fill a space completely and efficiently since the hexagons fill space without overlap. Moreover, Doehlert design is distinguished by a low ratio between the number of coefficients and the number of experiments. Thus, it can be considered as more efficient than the central composite design or Box-Behnken design [22]. Investigations of boron removal by Donnan dialysis were carried out considering four operating factors, i.e. the counterion concentration in the receiver compartment, the boron concentration, the pH of the feed compartment, and the type of the anionic-exchange membrane. The effect of these factors and their interactions were evaluated using the full factorial design. The Donnan dialysis was then optimized using the Response Surface Methodology based on the Doehlert design.

2. Methods and materials

2.1 Membranes

In the Donnan dialysis process, two membranes have been used, which are Neosepta® AFN and Neosepta® ACS (supplied by Alstom). Before any measurement, it is necessary to condition the samples to stabilize their physicochemical properties and to eliminate any impurities that could come from their manufacturing process. **Table 1** shows the properties of the used membranes determined according to the standard NF X 45-200 [23]. The water content was determined by the Mettler-Toledo moisture thermo balance device. The water content was calculated by Eq. (1):

$$WC(\%) = \frac{W_h - W_d}{W_h} \times 100 \tag{1}$$

where W_h is the mass of the hydrated membrane, Wd is the mass of the dried membrane, and WC(%) is the water content percentage. The water content is the difference in mass between the hydrated membrane (which has been immersed in the proper stabilization content and slightly compressed to remove the surplus liquid) and the dried membrane (which has been dried at 140°C until membrane mass stabilization indicates that all of the water has been removed).The mean value of 10 measurements at different locations using a 1−μm resolution Käfer Thickness Dial Gauge is the dry-state membrane thickness.

The ion-exchange capacity (in meq. of functional sites per gram of dry membrane or per cm^3 of wet membrane) was determined following the French standard NF X 45-200 [23].

In order to prepare the samples for the Donnan dialysis operations, the samples had to be conditioned before any measurement could be made. This was done primarily to eliminate contaminants from the production process and to stabilize their physical-chemical properties. French standard NF X 45-200 was followed in the conditioning process.

Membranes	ACS	AFN
Thickness (μm)	150	120
Water Content (%)	18.9	47.8
Ion-Exchange Capacity (meq/g)	1.85	3.00

Table 1.
Properties of the two membranes used.

2.2 Donnan dialysis (DD)

The Donnan dialysis is a method of membrane separation in which identically charged ions are exchanged between two solutions via an ion-exchange membrane [24–26]. The chemical potential gradient acts as the driving force in Donnan dialysis; anions are exchanged stoichiometrically through an anionic-exchange membrane, and the procedure is only complete if the Donnan equilibrium is attained. Since the electroneutrality is maintained, the feed should exchange the same number of anions with the receiver compartment in the opposite direction [19, 27, 28].

Figure 1 indicates the schematic flow of Donnan dialysis. The apparatus is used to study Donnan dialysis's removal of boron. It consists of a cell with feed and receiver compartments divided by an anion-exchange membrane inside a thermoregulated water bath. A peristaltic pump with two identical heads and a speed variator that allows for varied flow rates is used to pump the solutions through the cell. Through the use of two stirring rods with variable speeds, the hydrodynamic conditions on either side of the membrane can be changed. Two removable sections constructed of polymethylmetacrylate (plexiglass) make up the dialysis cell. It consists of four pieces that are connected by three threaded rods made of stainless steel. Supports provide the centering. The two tubes that make up the two compartments at the center are symmetrical. Each compartment has three threaded holes that provide support for inserting boxes. The membrane forms a seal by sandwiching itself between these two compartments. The feed and the receiver compartment were supplied, through a peristaltic pump, with NaCl and a containing boron solution, respectively. The used membranes were AFN and ACS. Boron concentrations were determined after 7 hours of treatment by Donnan dialysis for each experiment. By reacting the samples with azomethine-H and then measuring the absorbance at 420 nm with a UV-visible spectrophotometer, the samples' boron concentration was determined [29]. A linear concentration between 1 and 4 mg/L was found. Higher concentration samples were diluted to fit the previously mentioned linearity range.

Boron removal rate was determined by Eq. (2):

$$Y(\%) = \frac{C_0 - C_e}{C_0} \times 100 \tag{2}$$

With C_0 and C_e are the initial boron and equilibrium concentrations, respectively.

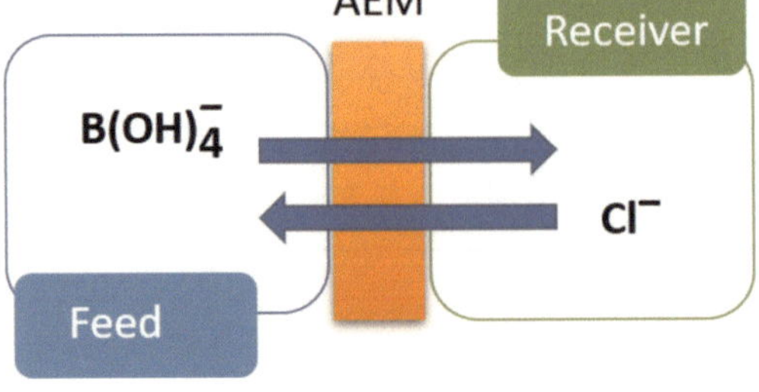

Figure 1.
Schematic flow of Donnan dialysis.

2.3 Optimization process

To identify the most significant and influencing parameter, a full factorial design was performed first, followed by an RSM design based on the Doehlert matrix. NemrodW® was the program employed in this study. The practical use of the Experimental Research Methodology (experimental designs) depends on the NemrodW® program.

3. The preliminary study

The definition of each factor's levels must be done with great care since if the domain is either too tiny or too large, the mathematical models might no longer work. Because of this, it is advised to do a preliminary study to help determine the appropriate high and low levels for each element. To establish the bounds of each parameter and better pinpoint the optimal value, a preliminary study was carried out as a phase in the parameters' pre-optimization process.

3.1 The pH effect in the feed compartment

The pH effect of the feed compartment is studied by varying the initial pH of the feed solution from 9.5 to 12.5. **Figure 2** shows the Boron removal rate under different initial pH values for a counterion concentration of 0.1 mol/L, an initial boron concentration of 50 mg/L, and a stirring speed of 500 rpm.

Due to its effect on the transfer of boron from the feed compartment to the receiver compartment, the effect of pH was first investigated. According to **Figure 2**, at a pH of 11.5 the AFN membrane removed the most boron (45%), while the ACS membrane removed 17%. This can be explained by the two species of boron that occur in aqueous solutions at various pH levels, which are the boric acid $B(OH)_3$ in diluted aqueous solutions below pH 7 and the metaborate anion $B(OH)_4^-$ at pH 10 [30, 31]. However, above a pH of 11.5, the presence and competition with OH are likely to have

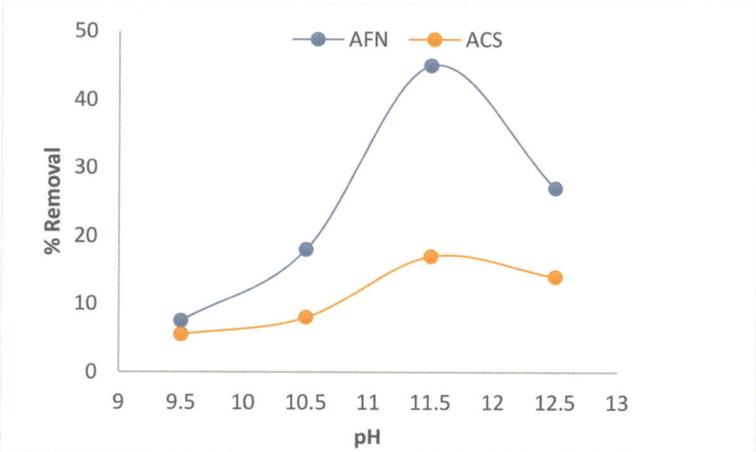

Figure 2.
Influence of pH on the removal rate of boron.

an impact on the transport of boron, which reduces boron removal because the hydroxyl ion transport is preferred because OH has much higher mobility than boron. In reality, the boron transfer process involves three steps. First, the boron in the feed solution is exchanged with ions or ionizable groups of the anion-exchange membrane. The second step involves the transport of boron across the membrane to the receiving solution side. In the third step, the boron is transported into the receiver solution following an exchange with the counterions to guarantee electroneutrality [32, 33]. Therefore, it can be said that for the two used membranes, the highest boron transport was attained at pH = 11.5 in this case.

3.2 Concentration of chloride in the receiver compartment effects performance

One of the factors influencing the elimination of boron through the anionic-exchange membrane during the Donnan dialysis procedure is the chloride concentration. To study the effect of this parameter, the concentration of the Cl counterion was varied from 0.01 mol/L to 0.05 mol/L in the receiver compartment for a boron concentration fixed at 50 mg/L. **Figure 3** displays the impacts of the two membranes' receiver compartment's chloride concentration.

According to **Figure 3**, the flux of boron ions through anion-exchange membranes increases with the increase of chloride concentration from 0.01 to 0.5 mol/L. At 0.01 mol/L of counterion, the boron removal rate reaches only 32% and 13% for AFN and ACS, respectively. At 0.5 mol/L of the concentration of Cl, the removal efficiency increases to 58% for AFN and 38% for ACS. The improvement in the cross-ion transfer between Cl and boron necessary to maintain electroneutrality is explained by the fact that the concentration gradient of the counterions grows.

For the two membranes, it appears that the improvement in the boron removal in the feed compartment, as shown by an improvement in the exchange's kinetics, is related to the increase in the concentration of counterions in the receiver compartment. In fact, it is known that ion exchange is faster when the concentration of counterions is higher in the receiver compartment [34–36].

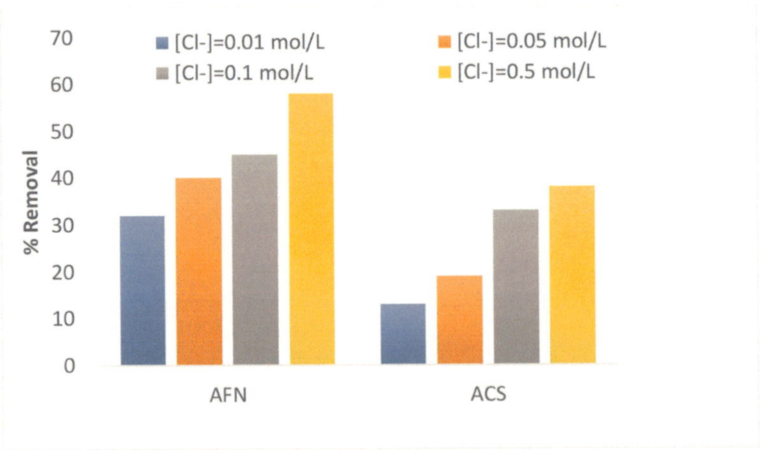

Figure 3.
Concentration of chloride in the receiver compartment effects performance.

3.3 Boron concentration effect

The elimination by Donnan dialysis is significantly influenced by the amount of boron in the feed compartment. Under these circumstances, the receiver compartment's Cl- concentration is 0.5 mol/L, the feed compartment's pH is 11.5, and the boron concentration rises from 5 mg/L to 100 mg/L. **Figure 4** presents the findings.

According to **Figure 4**, the improvement in cross-ion transfer between Cl^- and $B(OH)_4^-$ helps to maintain the gradient concentration of boron at a high level. The elimination was 40% for AFN and 17% for ACS at the feed compartment's lowest boron content (25 mg/L). When the boron content is raised to 100 mg/L, their elimination is improved to 75% with AFN and to 48% with ACS. A similar result was reported by Tor [37].

3.4 Membranes choice

According to the results of **Figures 3** and **4**, the boron removal rate by Donnan dialysis depends significantly on the AEM properties. Therefore, the most effective membrane is AFN since it allows to reach 75% after 7 hours of treatment against 48% for ACS membrane. According to Akretche, (i) a high exchange capacity boosts the selectivity between monovalent and multivalent anions due to the higher repulsion charge, (ii) a high thickness reduces diffusion, resulting in a lower ions flux, and (iii) a high water content can reduce permselectivity and encourage the penetration of bulky ions [38]. In actuality, the AFN membrane exhibits the largest water content, the highest ion-exchange capacity, and a higher permeability to monovalent than bivalent anions. On the other hand, the ACS has the lowest permeability due to its large thickness and low water content. The works of Ayyildiz [31], who reported that the removal of boron by Donnan dialysis is more successful using the AFN membrane, support this conclusion. AFN membrane has been chosen as the subject of the following investigation.

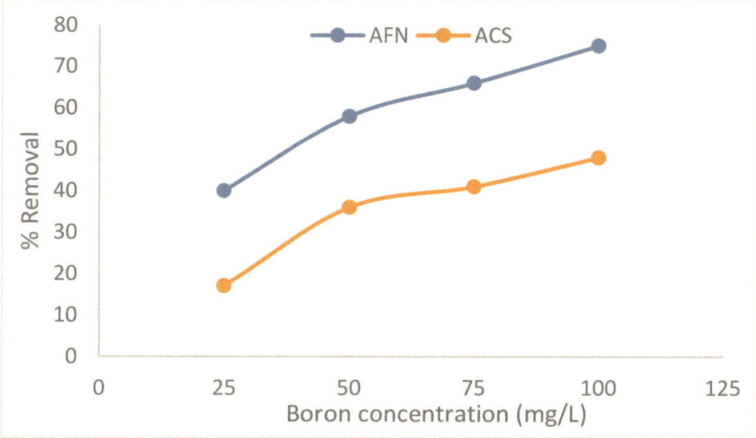

Figure 4.
Boron concentration effect.

4. Full factorial design

To ascertain the impact of these variables and how they interact with one another on the removal of boron by Donnan dialysis, the full factorial design was used. We were able to establish the experimental field and the level that had to account for every element thanks to the preliminary study. The starting boron concentration, counterion concentration, and feed compartment pH were the three variables that were selected. In order to more clearly define the examined response (boron removal efficiency), restrictions are imposed. The membrane AFN was used in the Donnan dialysis procedure.

A full factorial design was used to assess how operating parameters affected the removal of boron by Donnan dialysis. **Table 2** displays the experimental ranges and factors level. For each of the three component designs specified in the study, a full factorial matrix made up of eight distinct experiments was used. A linear polynomial model with interaction is used to model the experimental response related to a factorial design (see Eq. (3)):

$$Y = b_0 + b_1X_1 + b_2X_2 + b_3X_3 + b_{12}X_1X_2 + b_{13}X_1X_3 + b_{23}X_2X_3 \tag{3}$$

where Y is the experimental response, X_i denotes the coded variable, bi denotes the estimation of factor i's principal effect on the response Y, and b_{ij} denotes the estimation of factor i and factor j's interaction effect on the response Y.

The coefficients of the model were estimated in accordance with the findings in **Table 3**, and it was discovered that (see Eq. (4)):

Factors	**Symbol**	**Range and levels**	
Coded variable X_1	[B]	−1	1
Concentration of boron (mol/L)		25	100
Coded variable X_2	$[Cl^-]$	−1	1
Concentration of counter-ion (mg/L)		0.1	0.5
Coded variable X_3	pH	−1	1
pH of solution		10.5	12.5

Table 2.
Experimental range and factors level studied in the factorial design.

No	X_1	X_2	X_3	Y_B**(%)exp**	Y_B**(%)cal**
1	−1	−1	−1	19.9	20.2
2	+1	−1	−1	27.7	27.4
3	−1	+1	−1	30.7	30.4
4	+1	+1	−1	36.5	36.8
5	−1	−1	+1	36.9	36.6
6	+1	−1	+1	39.9	40.2
7	−1	+1	+1	41.9	42.2
8	+1	+1	+1	45.3	45.0

Table 3.
Full factorial design matrix.

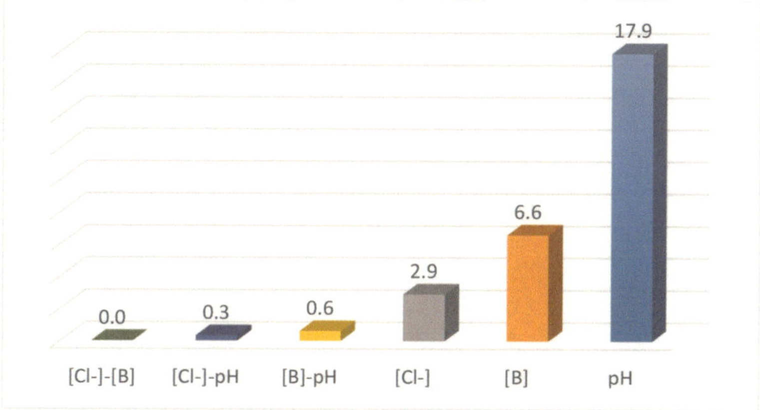

Figure 5.
Pareto analysis of the removal of boron.

$$Y(\%) = 34.85 + 2.50\ X_1 + 3.75\ X_2 + 6.15\ X_3 - 0.20\ X_1X_2 - 0.90\ X_1X_3 - 1.15\ X_2X_3 \quad (4)$$

The different coefficients of the polynomial model (Eq. (4); $R^2 = 0.999$) were determined, which represented the effects and interactions of the various investigated factors. The Pareto analysis (**Figure 5**) allows to evaluate the contribution of each parameter on the response according to the equation (Eq. (5)):

$$P_i = \left(\frac{b_i^2}{\sum b_i^2}\right)^2 \times 100 \quad (5)$$

The three investigated factors have a favorable impact on the observed behavior, i.e., increasing them improved boron removal. In contrast to the 17.9% they contributed to pH, and their contributions to the examined response were just 6.6% for boron concentration and 2.9% for chloride concentration. Thus, two factors—pH and boron concentration—can have a significant impact on the elimination. The solution pH coefficient's positive value indicates that boron removal was enhanced. This is brought on by the existence of, which, high pH levels, becomes the dominant species. The elimination of boron by Donnan dialysis is moderately affected by both the counterion concentration and the boron concentration. Therefore, the feed compartment's pH is the most crucial variable.

5. Doehlert design

In this investigation, the optimum condition was found using the Response Surface Methodology (RSM) in accordance with the Doehlert design. By evenly dispersing the experimental points within the space-filling of the variables, Doehlert's method is created. $N = k^2 + k + 1$ is the total number of experiments for k factors. Fifteen tests were in total, three of which were replicated in the central field [39, 40]. The initial boron concentration, feed compartment pH, and receiver compartment counterion

Factors	Coded symbol	Range and levels		
		−1	0	+1
Concentration of counter-ion (mol/L)	X_1	0.1	0.3	0.5
pH of the solution	X_2	10.5	11.5	12.5
Initial concentration of boron (mg/L)	X_3	25	62	100

Table 4.
Experimental range and levels of the factors.

concentration were all factors that were examined. According to the preliminary study, these parameters' upper and lower limits were established. To get the most information out of the system, it is typically preferable to use the variable with the significant effect as the variable with the highest levels. The experimental field of the factors under investigation is shown in **Table 4**.

An experiment conducted in the Doehlert domain is able to predict, at any point in the experimental domain, the value of an answer by estimating the coefficients of a second-order function [40]. The selected model uses a polynomial equation (see Eq. (6)) to describe the predicted values of the responses Y:

$$Y = b_0 + b_1X_1 + b_2X_2 + b_3X_3 + b_{11}X_1^2 + b_{22}X_2^2 + b_{33}X_3^2 + b_{12}X_1X_2 + b_{13}X_1X_3 + b_{23}X_2X_3 \quad (6)$$

The estimated principal effect of factor i is denoted by the letters b_i, the estimated second-order effects by the letters b_{ii}, the estimated interactions between factors i and j by the letters b_{ij}, and the coded variable by the letters X_i.

The regression coefficients (R^2) and the percentage absolute errors of deviations (AED) between experimental and calculated results must be considered for model validation. The AED was calculated from Eq. (7):

$$AED\ (\%) = \frac{100}{N} \cdot \left| \frac{Y_{exp} - Y_{theo}}{Y_{exp}} \right| \quad (7)$$

where Y_{exp} and Y_{theo} are the responses obtained from experiments and from the model, respectively. N is the number of points at which measurements were carried out. A model was considered valid if $R^2 > 0.7$ and AED $< 10\%$ [41].

For boron removal optimization, Response Surface Methodology via Doehlert design was used with membrane AFN. To optimize the factors, 15 experiments involving three variables were evaluated using a Doehlert experimental design (**Table 5**).

No	X_1	X_2	X_3	$Y(\%)_{Exp}$	$Y(\%)_{Cal}$
1	1.0	0.000	0.000	87.3	87.4
2	−1.0	0.000	0.000	81.5	81.4
3	0.5	0.866	0.000	51.4	51.6
4	−0.5	−0.866	0.000	16.1	15.9
5	0.5	−0.866	0.000	21.1	20.7

No	X_1	X_2	X_3	$Y(\%)_{Exp}$	$Y(\%)_{Cal}$
6	−0.5	0.866	0.000	50.7	50.6
7	0.5	0.287	0.816	65.2	64.9
8	−0.5	−0.287	−0.816	35.9	36.2
9	0.5	−0.287	−0.816	40.1	39.9
10	0.0	0.577	−0.816	46.6	46.5
11	−0.5	0.287	0.816	62.6	62.7
12	0.0	−0.577	0.816	39.5	39.6
13	0.0	0.000	0.000	84.2	84.2
14	0.0	0.000	0.000	84.2	84.2
15	0.0	0.000	0.000	84.2	84.2

Table 5.
Doehlert Matrix and obtained results.

Using the experimental results from **Table 5**, the second-order polynomial equation was fitted to the data appropriately and the coefficients were presented in Eq. (8):

$$\begin{aligned} Y = {} & 84.2 + 3.01X_1 + 18.81X_2 + 9.12X_3 + 0.20X_1^2 - 65.90X_2^2 - 37.40X_3^2 \\ & - 2.38b_{12}X_1X_2 - 010\ X_1X_3 - 12.20X_2X_3 \end{aligned} \quad (8)$$

Based on the obtained results, the coefficients show that pH of the feed compartment had a significant effect on boron removal (b_2 = 18.81). As a second influencing factor, boron concentration (b_3 = 9.12) was taken into consideration. However, chloride concentration had a less significant effect on boron removal (b_1 = 3.11). The feed compartment's pH and boron concentration (b23 = −12.20) had the most significant interaction and had an adverse impact on Donnan dialysis's ability to remove boron.

Although they had a small impact on the removal of boron by Donnan dialysis, the interactions between the chloride concentration and the feed compartment's pH (b_{12} = −2.38) and the concentration and boron concentration (b_{13} = −0.10) were not.

The regression coefficient (R^2) and the percentage of absolute errors of deviation (AED) were used to evaluate the model's validity. The regression coefficient (R^2) is better than 0.7, and the percentage of absolute errors of deviation (AED) (%) = 0.425% was less than 10%.

The contour plots (curve of constant response) are used to describe how boron is removed by Donnan dialysis. The response surface is represented in a contour plot as a two-dimensional plane where all points with the same response are joined to form contour lines with constant responses. Typically, a surface plot shows a three-dimensional image that could give a clearer idea of the response. To show the relationship between two factors and a response, use contour plots. The graph shows values of the pH for combinations of the $[Cl^-]$ and $B(OH)_4^-$. The $[Cl^-]$ and $B(OH)_4^-$ values are displayed along the X- and Y-axes, while contour lines and bands represent the response value Y. The obtained plots are illustrated in **Figure 6**.

At a constant boron content of 62 mg/L, the first plot displays the combined fluctuation of chloride concentration and pH. The contour plots' form demonstrates that, only at pH 11.5 when chloride concentration increases from 0.1 mg/L to 0.3 mg/L,

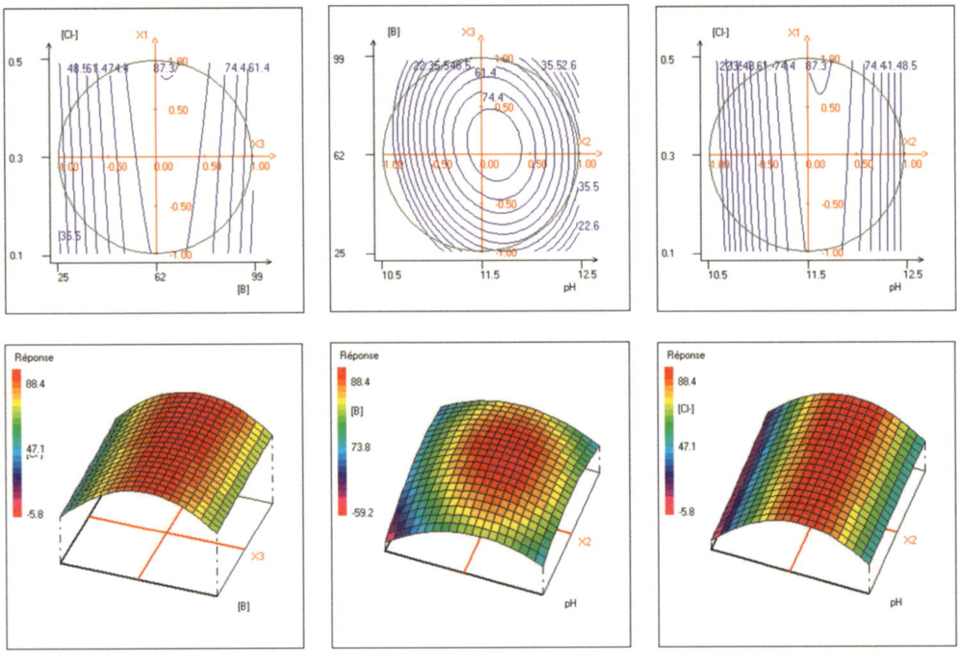

Figure 6.
Contour plots and three dimensions plots.

boron removal improves. This was attributable to the many boron forms that can exist in aqueous solutions at various pH levels. At higher pH levels, the $B(OH)_4^-$ is the dominating species. However, beyond pH 11.5, the presence and competition with OH-, which reduces the removal of boron, are likely to have an impact on the transport of boron. Therefore, it can be said that pH 11.5 produced the largest boron transfer.

The variation in boron concentration and pH at a fixed chloride concentration of 0.3 mol/L is shown in the second plot. The shape of these iso-response curves, which are concentrated in the center of the domain, demonstrates that the pH has a significant impact on the elimination of boron. This was anticipated because boron removal was positively influenced by the pH and chloride content. The variation of chloride and boron concentrations with a constant pH of 11.5 is depicted in the third plot. According to the contour plots, the rate of boron removal improves only when the chloride concentration is increased from 0.1 mg/L to 0.3 mg/L around the boron concentration of 62 mg/L.

The optimum values are 66 mg/L for the concentration of boron, 0.5 mol/L for the concentration of chloride, and 11.6 for the pH of the feed compartment. These conditions led to a maximum removal of boron of 88.8%. A replicate three times of experiment was conducted in the optimum conditions in order to verify the efficiency of predicting values. The coefficient of repeatability is less than 1%, so it can be concluded that the removal of boron by Donnan dialysis is reproducible.

6. Conclusion

The influence of operating parameters on boron removal by Donnan dialysis was investigated using two different membranes, AFN and ACS. For an initial boron concentration of 100 mg/L, a pH of 11.5 in the feed compartment and 0.5 mol/L of Cl-, the counterion, in the receiver compartment, 75% of boron removal rates for AFN and 48% for ACS were recorded. The influence and interactions of these parameters were then evaluated using a full factorial design. It was concluded that pH is the most influent in the elimination of boron. The Response Surface Methodology by Doehlert enabled the identification of the ideal working conditions for the removal of boron, which reached an efficiency of 88.8% using an AFN membrane, which were [B] = 66 mg/L, pH = 11.6 and [Cl-] = 0.5 mol/L. Compared to the conventional "one-at-a-time" approach, using the Response Surface Methodology to identify the optimal conditions for 13.8% can be seen of as a good option.

Acknowledgements

The researchers would like to thank the ICMPE/University Paris Créteil Est for funding the publication of this project.

Author details

Ikhlass Marzouk Trifi[1]*, Lasâad Dammak[2], Lassaad Baklouti[3] and Béchir Hamrouni[1]

1 Desalination and Water Treatment Research Laboratory, Faculty of Sciences of Tunis, University of Tunis El Manar, Tunis, Tunisia

2 Institute of Chemistry and Materials Paris-Est (ICMPE), Paris-Est University, Thiais, France

3 Department of Chemistry, College of Sciences and Arts at Al Rass, Qassim University, Saudi Arabia

*Address all correspondence to: ikhlassmarzouk@gmail.com

References

[1] Brdar-Jokanovic M. Boron toxicity and deficiency in agricultural plants. International Journal of Molecular Sciences. 2020;**21**:1424-1444. DOI: 10.3390/ijms21041424

[2] Pandey A, Khan MK, Hakki EE, Gezgin S, Hamurcu M. Combined boron toxicity and salinity stress—An insight into its interaction in plants. Plants. 2019;**8**:364-386. DOI: 10.3390/plants8100364

[3] Guler E, Kaya C, Kabay N, Arda M. Boron removal from seawater: State of the art review. Desalination. 2015;**356**: 85-93. DOI: 10.1016/j.desal.2014.10.009

[4] Chruszcz-Lipska K, Winid B, Madalska GA, Macuda J, Lukanko L. High content of boron in curative water: From the Spa to industrial recovery of borates? (Poland as a case study). Minerals. 2021;**11**:8-26. DOI: 10.3390/min11010008

[5] Lyu J, Zhang N, Liu H, Zeng Z, Zhang J, Bai P, et al. Adsorptive removal of boron by zeolitic imidazolate framework: Kinetics, isotherms, thermodynamics, mechanism and recycling. Separation and Purification Technology. 2017;**187**:67-75. DOI: 10.1016/j.seppur.2017.05.059

[6] Ting TM, Nasef MM, Aravindan D, Rosslan IFN, Ruslan N. Selective removal of boron from industrial wastewater containing high concentration of ammonia by radiation grafted fibrous adsorbent in fixed bed column. Journal of Environmental Chemical Engineering. 2021;**9**:104993. DOI: 10.1016/j.jece.2020.104993

[7] Yilmaz A, Boncukcuoğlu R, Kocakerim M. A quantitative comparison between electrocoagulation and chemical coagulation for boron removal from boron-containing solution. Journal of Hazardous Materials. 2007; **149**:475-481. DOI: 10.1016/j.jhazmat.2007.04.018

[8] Chen M, Dollar O, Shafer-Peltier K, Randtke S, Waseem S, Peltier E. Boron removal by electrocoagulation: Removal mechanism, adsorption models and factors influencing removal. Water Research. 2020;**170**:115362. DOI: 10.1016/j.watres.2019.115362

[9] Turek M, Dydo P, Bandura-Zalska B. Boron removal from dual-staged seawater nanofiltration permeate by electrodialysis. Desalination Water Treatment. 2009;**10**:60-63. DOI: 10.5004/dwt.2009.782

[10] Guesmi F, Louati I, Hannachi C, Hamrouni B. Optimization of boron removal from water by electrodialysis using response surface methodology. Water Science and Technology. 2020;**81**: 293-300. DOI: 10.2166/wst.2020.105

[11] Ali Z, Al Sunbul Y, Pacheco F, Ogieglo W, Wang Y, Genduso G, et al. Defect-free highly selective polyamide thin-film composite membranes for desalination and boron removal. Journal of Membrane Science. 2019;**578**:85-94. DOI: 10.1016/j.memsci.2019.02.032

[12] Li Y, Wang S, Song X, Zhou Y, Shen H, Cao X, et al. High boron removal polyamide reverse osmosis membranes by swelling induced embedding of a sulfonyl molecular plug. Journal of Membrane Science. 2020;**597**:117716. DOI: 10.1016/j.memsci.2019.117716

[13] Kumar R, Ahmed M, Ok S, Garudachari B, Thomas JP. Boron selective thin film composite

nanofiltration membrane fabricated via a self-assembled trimesic acid layer at a liquid–liquid interface on an ultrafiltration support. New Journal of Chemistry. 2019;**43**:3874-3883. DOI: 10.1039/C8NJ05670F

[14] Lan N, Wang KY, Weber M, Maletzko C, Chung TS. Investigation of novel molecularly tunable thin-film nanocomposite nanofiltration hollow fiber membranes for boron removal. Journal of Membrane Science. 2021; **620**:118887. DOI: 10.1016/j.memsci.2020.118887

[15] Alharati A, Swesi Y, Fiaty K, Charcosset C. Boron removal in water using a hybrid membrane process of ion exchange resin and microfiltration without continuous resin addition. Journal of Water Process Engineering. 2017;**17**:32-39. DOI: 10.1016/j.jwpe.2017.03.002

[16] Hussain A, Sharma R, Minier-Matar J, Hirani Z, Adhama S. Application of emerging ion exchange resin for boron removal from saline groundwater. Journal of Water Process Engineering. 2019;**32**:100906-100915. DOI: 10.1016/j.jwpe.2019.100906

[17] Eryildiz B, Yuksekdag A, Korkut S, Koyuncu İ. Performance evaluation of boron removal from wastewater containing high boron content according to operating parameters by air gap membrane distillation. Environmental Technology & Innovation. 2021;**22**: 101493. DOI: 10.1016/j.eti.2021.101493

[18] Turek M, Bandura-Zalska B, Dydo P. Boron removal by Donnan dialysis. Desalination and Water Treatment. 2009;**10**:53-59. DOI: 10.5004/dwt.2009.711

[19] Bryjak M, Duraj I. Anion-exchange membranes for separation of borates by Donnan dialysis. Desalination. 2013;**310**: 39-42. DOI: 10.1016/j.desal.2012.07.025

[20] Freeth FA. Frederick George Donnan, 1870-1956. Biographical Memoirs of Fellows of the Royal Society. 1957;**3**. DOI: 10.1098/rsbm.1957.0003

[21] Marzouk I, Dammak L, Chaabane L, Hamrouni B. Optimization of chromium (VI) removal by Donnan dialysis. American Journal of Analytical Chemistry. 2013;**4**:306-313. DOI: 10.4236/ajac.2013.46039

[22] Marzouk Trifi I, Trifi B, Djamel A, Hamrouni B. Simultaneous removal of nitrate and nitrite by Donnan Dialysis. Environmental Engineering and Management Journal. 2020;**20**(6): 973-983

[23] French Standard NF X 45-200. Membranes Polymers Échangeuse d'Ions. Paris, France: AFNOR; 1995

[24] Strathmann H. Ion-Exchange Membrane Separation Processes. 1st ed. Amsterdam: Elsevier Science; 2004

[25] Nouri S, Dammak L, Bulvestre G, Auclair B. Studies of the crossed ionic fluxes through a cation-exchange membrane in the case of Donnan dialysis. Desalination. 2002;**148**:383-388. DOI: 10.1016/S0011-9164(02)00734-8

[26] Mokrani S, Dammak L, Bulvestre G, Auclair B. Experimental and theoretical studies of the crossed ionic fluxes through a cation-exchange membrane. Journal of Membrane Science. 2002;**199**: 147-160. DOI: 10.1016/S0376-7388(01) 00692-5

[27] Wisniewski J, Rozanska A. Donnan dialysis with anion exchange membranes as a pretreatment step before electrodialytic desalination.

Desalination. 2006;**191**:210-218. DOI: 10.1016/j.desal.2005.06.038

[28] Rozanska A, Wiśniewski J, Winnicki T. Donnan dialysis with anion-exchange membranes in a water desalination system. Desalination. 2006;**198**:236-246. DOI: 10.1016/j.desal.2006.02.006

[29] Rodier J. L'analyse de l'eau. 9th ed. Paris: Edition Dunod; 2009

[30] Zohdi N, Mahdavi F, Abdullah LC, Choong TSY. Removal of boron from aqueous solution using magnetic carbon nanotube improved with tartaric acid. Journal of Environmental Health Science and Engineering. 2014;**12**:3-15. DOI: 10.1186/2052-336X-12-3

[31] Ayyildiz HF, Kara H. Boron removal by ion exchange membranes. Desalination. 2005;**180**:99-108. DOI: 10.1016/j.desal.2004.12.031

[32] Sato K. Effects of the feed solution concentration on the separation degree in Donnan dialysis for binary systems of amino acids. Journal of Membrane Science. 2002;**196**:211-220. DOI: 10.1016/S0376-7388(01)00598-1

[33] Hichour M, Persin F, Sandeaux J, Gavach C. Fluoride removal from waters by Donnan dialysis. Separation and Purification Technology. 2000;**18**:1-11. DOI: 10.1016/S1383-5866(99)00042-8

[34] Veerman J. The effect of the NaCl bulk concentration on the resistance of ion exchange membranes—Measuring and modeling. Energies. 2020;**13**:1946-1972. DOI: 10.3390/en13081946

[35] Amang DN, Alexandrova S, Schaetzel P. Mass transfer characterization of Donnan dialysis in a bi-ionic chloride–nitrate system. Chemical Engineering Journal. 2004;**99**: 69-76. DOI: 10.1016/j.cej.2003.11.030

[36] Wisniewski J, Rozanska A, Winnicki T. Removal of troublesome anions from water by means of Donnan dialysis. Desalination. 2005;**182**:339-346. DOI: 10.1016/j.desal.2005.02.032

[37] Tor A, Çengeloğlu Y, Ersöz M, Arslan G. Transport of chromium through cation-exchange membranes by Donnan dialysis in the presence of some metals of different valences. Desalination. 2004;**170**:151-159. DOI: 10.1016/j.desal.2004.04.002

[38] Akretche D, Kerdjoudj H. Donnan dialysis of copper, gold and silver cyanides with various anion exchange membranes. Talanta. 2000;**51**:281-289. DOI: 10.1016/S0039-9140(99)00261-1

[39] Hosni N, Zehani K, Djebali K, Mazaleyrat F, Bessais L, Maghraoui-Meherzi H. Experimental design approach for the synthesis of 3D-CoFe2O4 nanoflowers thin films by low-cost process. Materials Chemistry and Physics. 2020;**255**:123493-123503. DOI: 10.1016/j.matchemphys.2020.123493

[40] Trifi B, Marzouk Trifi I, Kouass S, Zahraa O, Alatrache A. Gliding arc plasma treatment of levofloxacin using experimental design approach. Clean - Soil, Air, Water. 2023;**00**:1-13. DOI: 10.1002/clen.202200364

[41] Nde Bup D, Abi CF, Tenin D, Kapseu C, Tchiegang C. Optimisation of the cooking process of sheanut kernels (vitellaria paradoxa gaertn.) using the Doehlert experimental design. Food and Bioprocess Technology. 2012;**5**:108-117. DOI: 10.1007/s11947-009-0274-z

Chapter 3

Boron Nutrition in Horticultural Crops: Constraint Diagnosis and Their Management

Pauline Alila

Abstract

Out of 30 elements, 16 horticulture crops require them to thrive. All horticultural crops require boron, one of the necessary elements, to function. Extremely trace amounts of boron are present in soils. The majority of the boron that is readily available in humid areas is primarily contained in organic debris, which is broken down by microorganisms for the benefit of plants. In the tropics this element is leached down in soil due to heavy rainfall. As trace element B plays an important role in the growth and development of plants. Various crops exhibit symptoms of deficiency as well as of toxicity when there is even a slight aberration of available boron in soil. Therefore, it is imperative to study and understand the optimum requirement of B by specific crops. Boron also interacts with other elements and manifests in crop plants in various ways. This chapter attempts to understand some of the roles of boron in horticultural crops (fruits and vegetables) and its management for optimum growth and development in crop plants.

Keywords: boron, its importance, diagnosis, horticultural crops, management

1. Introduction

Among the essential elements, boron plays an important role in all the horticultural crops. The microbial decomposition of organic matter in soil releases the available boron for use in plants particularly in humid region. Most of the different macro and micronutrients become exhausted in soil due to continuous uptake of these elements by the crops, thereby causing deficiency in plants which manifests in nutritional disorders often resulting in low yields. Warington in the year 1923 [1] demonstrated the essentiality of boron (B) for the growth and development of higher plants. Boron is crucial as a vital micronutrient for achieving optimal crop growth, development, yield, and quality. It plays a key role in forming and stabilizing cell walls, promoting lignification, and facilitating xylem differentiation [2]. Moreover, boron is essential for enhancing the protein and enzymatic functions of cell membranes, thereby improving membrane integrity. It's worth noting that sexual reproduction in plants is more vulnerable to boron deficiency compared to vegetative growth. During the pollination of flowers, boron supports pollen tube growth, ensuring successful seed set and fruit development. Additionally, boron contributes to increased nectar production in flowers, attracting pollinating insects. In the realm of crop production, boron stands out as a critical micronutrient required for the healthy growth of most crops. Furthermore, boron imparts drought tolerance to plants. Among the essential micronutrients, boron is unique as the sole non-metal present in a non-ionic form. Commercially, boron is sourced from minerals such as ulexite, natural boric acid (sassolite), borax (tincal), colemanite, and kernite, with the United States (US) and Turkey serving as the primary and richest sources of this element.

In vascular plants, known as tracheophytes, boron follows a path from the roots through the transpiration stream to the points of active growth, where it collects in the stems and leaves. It has been proposed that the localized accumulations of boron in these growing tissues have contributed to the evolutionary development of a reliance on boron for certain metabolic processes in plant meristems. Once boron accumulates in leaves, its subsequent movement within the plant is limited, and it becomes fixed in the apoplast. What sets boron apart from other nutrients is the considerable variation in its mobility among different plant species. The mobility of boron is facilitated by its interaction with polyols. In species where polyols (such as sorbitol, mannitol, and dulcitol, which are water-soluble sugar alcohols) are the primary products of photosynthesis, boron can move rapidly and effectively through the phloem. Examples of such species include those from the Malus, Prunus, and Pyrus genera (including apple, prune, plum, peach, pear, cherry, almond, and apricot), as well as olive and coffee in fruit plants, and carrot, onion, pea, celery, bean, cauliflower, cabbage, asparagus, wheat, and barley in vegetables and cereals. Typically, boron concentrations are highest at the tips and edges of leaves, but in species where boron is phloem-mobile, the concentration remains consistent throughout the leaves. Once again, it is suggested that boron's movement through the phloem is limited when boron levels are higher in older leaves. The symptoms of boron toxicity, which manifest at the edges of older leaves, have been interpreted as evidence of boron's immobility within plants [3]. Conversely, when there is a higher concentration of boron in young leaves and fruits, it serves as an indicator of phloem mobility [4–7]. Research indicates that maintaining optimal levels of boron enhances the germination of pollen and the growth of pollen tubes in almond trees, leading to improved fruit development and seed growth [8]. Applying boron as a foliar treatment to sour cherry trees just before the leaves fall significantly boosts boron levels in flower buds and enhances fruit set [9]. When boron is applied to trees in the autumn, it translocates from the leaves to the adjacent buds, where elevated levels persist and become evident in the flowers during anthesis. Generally, flowers receive their boron supply from reserves in the wood, which become mobilized during the development of the flowers. Boric acid, being a soluble compound susceptible to leaching by rainfall, frequently results in boron deficiency in regions with high rainfall and humidity. In contrast, soil boron toxicity is less common and tends to occur in arid and semi-arid areas. The concentration and availability of boron in soils are influenced by various factors, including the soil's parent material, texture, clay mineral composition, pH levels, liming, organic matter content, and its interactions with other elements. Consequently, understanding these factors that affect boron uptake is essential for assessing boron deficiency and toxicity under different environmental conditions.

2. Diagnosis of boron deficiency by soil and plant analysis

As early as 1942, Dregne and Power [10], reported that B availability in the soil is influenced by numerous factors such as soil reaction, soil moisture, active calcium and organic matter content. Boron deficiency in soil is mostly found in acidic soils, irrespective of parent material where excess B is leached out. Soils arising from igneous and metamorphic sandstone rocks are found to be naturally low in B. soils deficient in B are also found in laterite soils which have limited silica and strong Fe and Mn. Although alkaline soils are rich in total B but poor in available B for plants. Similarly, in peat soils and acid sludge and soil with low clay composition B deficiency is evident. In general, it is estimated that the total B concentration is between 20 and 200 mg B/kg, where its available concentrations vary greatly in different soil types. Boron is available to plants in water soluble form in soil. It is naturally present in the soil in the water-soluble form, which is readily available as a nutrient for plants. Movement of boron in soil is through "mass flow" which is absorbed by plants through transpiration stream as in water absorption. Consequently, in dry soils, boron uptake may be reduced or unavailable at a time of maximum B requirements in plants. Certain environmental conditions can also lead to low availability of soil boron. High rainfall can cause leaching of available boron from the root zone which is a major problem, especially in coarse-textured soils and when plants are at or just before the rapid growth of leaves and flower development stages. The other adverse environmental condition is when less rainfall or drought period occurs just prior to, or during flowering and seed set.

The requirements for boron (B) in plant growth exhibit significant variability across different plant species, as well as within the same species at various growth stages [2, 11–14]. In soil, a boron level below 0.5 mg/kg of water-soluble boron is considered low or deficient. An optimal or medium range of boron falls within the range of 0.5–2 mg/kg, while soils containing more than 2 mg/kg are deemed to have a high or excessive boron content. Consequently, soils containing less than 0.5 mg/kg of hot water-soluble boron are generally incapable of providing sufficient boron to support normal growth and yield in most plant species.

The total boron content in Indian soils has been observed to vary between 7 and 630 mg B/kg [15]. Reports of boron deficiency have primarily emerged from the soils in Indian states such as Assam, Bihar, Meghalaya, West Bengal, Jharkhand, and Odisha. This deficiency is notably prevalent in acid-red and lateritic soils, including high-pH calcareous soils [16]. Light-textured soils, like sandy loam and loamy sands, also tend to lack sufficient boron content due to their excellent drainage properties, which result in effective leaching [17]. Given that boron is a weakly held anion, it can be easily washed out of the soil, rendering acid-sand soils particularly susceptible to boron deficiency [18].

Leaf analysis, supplemented by soil analysis, can be reliable diagnostic tool for analysis of B availability and status in plants. The leaf analysis is derived from four factors:

- A leaf is a primary site for metabolic activities;
- Variations in nutrients are reflected in leaf content;
- The variations are more noticeable at certain developmental stages than others; and
- Nutrient levels in leaves at a particular growth stage are linked to crop productivity.

Boron, which plants absorb in the form of undissociated boric acid (B $(OH)_3$ or $H3BO_3$), possesses a strong capacity to create complexes within the plant system.

In the case of most crops, a boron content ranging from 15 to 100 mg/kg in plant tissue is considered sufficient for regular growth. Conversely, levels exceeding 200 mg/kg may be excessive, leading to a potentially toxic or inhibitory impact on crop growth and yields. When the concentration of boron in plant tissue falls below a critical threshold, it hinders the optimal growth and yield of the crop. Plants have quite specific demands for boron, and the gap between excess and deficiency is quite narrow. For instance, maintaining a leaf boron concentration within the range of 30–70 μg/g is ideal in mini cucumbers. Deficiency becomes evident when levels drop below 20 μg/g, while toxicity is observed when concentrations exceed 100 μg/g. Notably, boron does not readily migrate from older to newer plant tissues, making the roots to continuously uptake essential for the plant to grow normally. The importance of boron in pollination and in regulation of cells development is well established and its deficiency leads to poor seed set and fruit development. The cost of boron in correcting boron deficiency is justified with improvement in both the growth and yields of crop [19].

3. Interaction of boron with other elements

There has been reports of interactive effect of boron in plants with the other nutrient elements. In general, excessive applications of potassic fertilizers results in high potassium concentration which induces boron deficiency. B has been found to counter the toxic effects of aluminum on root growth of dicotyledonous plants [20]. In limestone soils rich in soil calcium reduces the availability of boron. Conversely, when there are excessive boron levels in soil, plant toxicity due to boron can be prevented by calcium application. Reports on the application of zinc to neutralize toxic effects of boron in some plants and there was subsequent increase in the crop yield is documented [19].

4. Correction of boron deficiency

When plants display deficiency symptoms of Boron, the issue can be easily addressed by administering borax, which is a white crystalline salt that is needed in very small amounts. Applying 5–10 kg/ha of boron can rectify boron deficiency in most plants. Boron can be delivered through both foliar and soil fertilization methods to ensure an adequate supply to plants. Foliar application is most effective when root activity is restricted and boron deficiency symptoms become evident during dry soil conditions in the growing season. Generally, a 0.2% borax solution can be sprayed onto the foliage every 10 days until the deficiency is corrected. For many vegetable crops, Solubor (100 g/100 L of water) can also be utilized as a foliar spray [19]. However, Das et al. [21] noted that soil application of boron outperforms foliar sprays. In calcareous soils, introducing sulfur at a rate of 2 MT/ha helps reduce soil pH to the range of 6.0–7.0 and enhances the boron solubility in the soil solution. To address hidden deficiencies, a spray of 0.2% boric acid or borax during the pre-flowering or flower head formation stages has been shown to increase yields. Typically, the application of boron ranges from 0.25 to 3.0 kg/ha, depending on the specific crop requirements and the chosen application method. Broadcast applications typically necessitate higher rates compared to banded soil applications or foliar sprays. It's crucial to exercise caution when addressing boron-deficient crops, as boron is highly toxic to most plants at relatively low levels. Additionally, ensuring an even distribution across the soil is essential to prevent crop plants from experiencing toxicity. In general, applying 10 kg of borax (Granubor) per hectare to deficient soil before planting is an effective preventive measure against boron deficiency.

5. Effect of boron in vegetable crops

Boron requirements vary among different plant species, and this variability may be attributed to differences in cell wall composition. Cereals and grasses exhibit lower sensitivity to low levels of available boron compared to legumes and certain vegetable crops. In dicots such as soybean and alfalfa, it's observed that the deficiency concentration of boron is three to four times higher in younger leaves than in older ones, which indicates the limited mobility of boron through the phloem in these species. Given boron's immobility within plants, soils with marginal boron levels can lead to crop deficiencies during critical growth phases, namely, during vegetative growth, flowering, and seed development. Consequently, maintaining a consistent boron supply during the growth season is crucial for achieving optimal growth and the seed yield. Experimental evidence demonstrates that boron significantly enhances yield, mobility, and stress tolerance in various crop species (see **Table 1**). In extreme cases, crops grown in low-boron soils perform well until the flowering stage, at which point there can be substantial yield losses due to floral abortion or failed seed set. The most effective time for boron application is at the beginning of the reproductive phase, considering its immobility within plants, and this timing depends on the photosynthetic efficiency of the specific plants. Some examples of deficiency symptoms are outlined in **Table 2**.

The issue of hollow stem in cauliflower was effectively addressed by applying Farm Yard Manure at a rate of 7.5 tons/ha as a basal application, coupled with either the use of Boric Acid at 0.3% or Liquid Boron at 1.5 g/l, carried out 30 days after planting. Notably, the application of Liquid Boron was found to be the more cost-effective choice. This boron application significantly resolved these problems and

Salient achievement	Reference
In sunflower, application of B @ 1.5 kg/ ha gave highest seed yield (2.01 t/ha)	[23]
In maize, mustard, onion, sweet potato, and sunflower, the application of B @ 1.5 kg/ha increased the B concentration and uptake.	[24]
In green gram B application increased the dry matter, yield and uptake of B	[25]
In groundnut, soil and foliar application of B had positive effect on growth and yield	[26]
Application of boronated super phosphate recorded significantly higher dry matter and yield of sunflower	[27]
Significant interaction effect of P × S × B was found for seed and oil yield of castor	[28]
Soybean grain yield was found to be statistically at par with either soil application of 20 kg sodium tetraborate (14% B) or two foliar sprays of 0.2% solution of the same salt.	[29]
Foliar application of B @ 0.2% at flowering stage of summer mungbean was found to be optimum for optimum economic yield in West Bengal.	[30]
Tuber number and yield of potato significantly increased with B fertilization.	[31]

Table 1.
Influence of B application on various crop plants [22].

Crop	Symptoms	Reference
Cauliflower	Roughened stems, leafstalks and midribs, poor flowering and development of curds with brown patches called Brown rot/ Browning, deformed and discolored brown heads with an empty space known as "Hollow Heart"	[19]
Cabbage	Distorted leaves and hollow areas in stems, cracked and corky stems, petioles, midribs and small heads.	[19]
Tomato	Stunted growth, dwarfed, leaves are twisted and small with variegated appearance. Yellowing and death of fruits in severe deficiency, fruits may be ridged, show corky patches, ripen unevenly and show cracking	[19]
Sugar beet	Young plants have distorted crown leaves. The leaf are small sized which becomes scorched and ultimately wilts	[19]
Turnip	Brown or gray concentric rings develop inside the root	[19]
Carrot	Leaves show longitudinal splitting, rough and small roots are formed which has a distinct central white core and top starts browning. The fruits also show blackened spots	[19]
Cucumber	Abnormal shoots, stunted apical growing points which eventually die. The new leaves are distorted and appear mottled while older leaves develop yellow border. Cracks in the stem and fruits are often deformed. The skin shows longitudinal and mottled yellow streaks leading to corky mark development on the fruits.	[32]
Muskmelon	Cracked stems and growing points dieback, leaves small, fruits distorted	[33]
Pepper	Stunted or dwarfed, leaves small, twisted and discolored, fruits twisted and show black spot	[32]
Bittergourd	Distortion of new leaves, appearance of broad yellow margins of oldest leaves	—
Okra	Leaves brittle and distorted, irregular lobe development	—
Solanum gilo	Upper leaves small, deformed and twisted, few flowers form, growing point die	—

Table 2.
Visible symptoms of boron deficiency in some vegetable crops.

led to an increase in curd yield [34]. In the case of muskmelons, applying boron improved both plant growth and fruit quality. Additionally, it reduced the occurrence of chilling injury in harvested fruit during cold storage at 5°C [33]. An optimal treatment combination consisting of 120 kg N/ha along with 0.6 kg B/ha was identified for achieving the maximum yield of tomatoes [35]. Furthermore, Naz et al. [36] reported that applying 2 kg B/ha resulted in the highest number of flower clusters per plant, fruit set percentage, total yield, fruit weight retention, and total soluble solids for the Rio Grand and Rio Fique tomato cultivars.

In a four-year study assessing the direct and residual effects of applied boron (B) in French bean-cauliflower cropping sequences, it was observed that applying 8 kg/ha of B led to reduced crop yields in eight crops spanning 4 years. When 2 kg/ha of B was applied, it enhanced French bean yields in soils with low-available B, whereas high-available B soils initially reduced yields in the first two crops but improved yields in the third and fourth crops. However, in both French beans and cabbage, elevated levels of B in plant tissue resulted in toxicity. The hot water-soluble B (HWS-B) content at harvest for each crop indicated a rapid decrease in the availability of applied B in these soils. Applying high levels of B-containing fertilizer to these soils led to the accumulation of B in toxic concentrations, resulting in lower vegetable yields.

For French beans, which are low accumulators of B and more sensitive crops, it is advisable to rely on residual B rather than applying B-containing fertilizer directly to the soil in any vegetable cropping system, especially in red soils [37].

Solanki et al. [38] found that the application of boron at rates of up to 1 kg B/ha led to a significant increase in the yields and dry matter production of vegetable crops, including carrot, cauliflower, and onion. However, when the boron level was raised to 2 kg B/ha, yields tended to decline. The extent of this response varied from one crop to another and followed a descending order of magnitude: carrot > cauliflower > onion. Additionally, the application of boron resulted in an improvement in both the content and yield of protein in vegetable crops compared to the control. Over time, the introduction of boron into the soil led to a progressive increase in its concentration and uptake by vegetable crops. The highest removal of boron was observed in cauliflower curds, while the lowest was noted in carrot roots. The percentage of apparent boron recovery was influenced by its levels, with the highest recovery occurring at the 1 kg B/ha level. However, as boron levels increased, boron use efficiency decreased, with the lowest efficiency recorded at 1.5 kg B/ha.

6. Effect of boron on fruit crops

Boron has been identified as a crucial element influencing reproductive processes, impacting anther development, pollen germination, and pollen tube growth. Apple trees, scientifically known as *Malus domestica* Borkh., are particularly recognized for their high demand for boron [14]. The initial visual indicators of boron deficiency are evident in poor fruit set, which subsequently leads to reduced yields. This deficiency is critical because boron plays a pivotal role in reproductive growth [39]. In apple trees suffering from boron deficiency, the fruits tend to be undersized, corky, distorted, and prone to cracking and russeting. They exhibit yellow skin and fail to develop a healthy red coloration [40]. Additionally, apple fruits may have diminished concentrations of soluble solids and acids when boron levels are insufficient [41]. Stone fruits are likewise affected by boron deficiency. For instance, cherry shoots that lack adequate boron growth initially but subsequently undergo necrosis at the tips and ultimately die. In B-deficient plants, some buds may remain closed during springtime, while others wither and perish. The presence of cracking, deformation, shriveling, both internal and external browning, as well as corking around the pit and in the flesh, serves as clear indicators of boron deficiency in cherry fruit [42]. Similar responses to low-boron soils are also observed in nut crops grown in temperate regions.

Boron stands as an indispensable mineral for the growth and proper functioning of citrus plants; however, it is frequently deficient in many types of soil. Pronounced deficiency, ranging from 39 to 68%, is notably prevalent in red and lateritic soils, as well as in leached acidic soils within the hot semi-arid ecoregion. It is also observed in soils derived from alluvium within the hot sub-humid ecoregion, brown and red hill soils in the warm, humid ecoregion, and highly calcareous soils within the hot sub-humid ecoregion. Interestingly, these are the specific climate and soil combinations that yield abundant high-quality citrus crops in countries like Brazil, China, and Japan. Ruchal et al. [43] reported an increased fruit set in mandarin trees in response to higher micronutrient concentrations in their application. This could possibly be attributed to the enhanced translocation of essential nutrients and hormones to the ovary tissue, thereby stimulating fruit formation. Another potential factor contributing to this effect could be the improved availability of microelements, which enhances photosynthesis, reduces fruit drop, and results in improvements in fruit size and quality. Assam lemon (*Citrus limon* (L.) Burm.), an indigenous lemon cultivar of Assam, is extensively grown on the warm southern slopes of the Himalayas in north-eastern India. This particular lemon variety is characterized by its ability to bear fruits in multiple flushes throughout the year, necessitating adequate nutrition to achieve optimal yields with high-quality fruits. Sheikh et al. [44] noted that the treatment involving ZnSO4 (0.2%), FeSO4 (0.2%), Borax (0.2%), and CuSO4 (0.2%) led to improved lemon fruit yields and quality. This treatment also resulted in the highest number of fruits per plant (73), yield per plant (11.5 kg), fruit fresh weight (158 g), fruit length (9.60 cm), fruit diameter (5.80 cm), juice content (152 mL/fruit), TSS (6.40°B), ascorbic acid content (49.10 mg/100 g), total sugar (6.30%), reducing sugar (3.90%), non-reducing sugar (2.40%), and the lowest titratable acidity (3.13%).

Guava trees that were treated with boric acid at concentrations of 0.1% and 0.2% exhibited notable improvements in various aspects of growth and fruit development. These improvements included increased extension of the terminal shoot, greater leaf count, enhanced leaf area per shoot, and accelerated fruit ripening, with reductions of 7 and 11 days, respectively. Additionally, the yield increased by 82 and 73%, respectively, with the higher concentration producing a slightly lower yield increment. Furthermore, the fruit size saw an increase when treated with 0.1% boric acid [45]. Pre-flowering applications of boric acid at concentrations of 0.1, 0.2, 0.3, or 0.4% on Allahabad Safeda guava resulted in significant enhancements in growth, flowering, and fruiting processes [46]. For Guava cv. L-49, the largest fruits, measuring 6.68 × 7.12 cm and weighing 125.8 g, were obtained with the application of 3.0% urea and 0.3% boric acid [47]. Similarly, spraying borax at a concentration of 0.2% effectively increased the size of Sardar guava fruits, their weight (95.25 g), and the yield (63.49 kg/tree) [48]. Combining the spray of borax at 0.2% with urea at 2% in three applications (pre-flowering, fruit setting, and 3 weeks after fruit setting) for Allahabad Safeda guava resulted in fruits measuring 4.84 cm in length, 5.00 cm in width, weighing 72.67 g, and yielding 19.08 kg per tree. Alternatively, foliar applications of borax at 0.2% alone recorded a higher fruit weight of 80.67 g and a yield of 20.17 kg per plant [49]. Yadav [50] observed that the best yield of high-quality fruits (67.7 kg per tree), the highest number of fruits per tree (686), and the largest fruit volume (107.5 cc) were achieved with foliar application of a combination of urea (3.0%), borax (0.15%), and NAA (10 ppm) in guava trees. Furthermore, a pre-harvest spray of borax with concentrations ranging from 0.2 to 1.2% applied twice in October resulted in improved guava fruit quality in the Sardar cultivar, particularly in terms of size and weight. In another experiment involving foliar application of H3BO3 at concentrations of 0.3, 0.5, and 1.0% on guava cultivar L-49 during the winter season [51], fruit weight and yield both increased, with the highest values recorded at 1.0% B, reaching 141.0 g and 73.0 kg/tree, respectively.

The most effective method for reducing fruit drop in rose-scented litchi, as reported by Ref. [52], was the application of borax. Through foliar sprays of borax at concentrations of 0.5 and 1.0%, fruit drop was reduced to 75–76%, compared to the 92.4% fruit drop observed in the control group. In the case of litchi, Haq and Rab [53] found that foliar application of CaCl2 and borax led to significant increases in the average fruit skin calcium content (from 4.79 mg/100 g DW in the control to the highest 8.88 mg/100 g DW with CaCl2 3% + boron 1.5% treatment), boron content (from 0.109 mg/100 g DW in the control to 0.247 mg/100 g DW with the same treatment), and skin strength (from 2.43 kg/cm2 in the control to 3.01 kg/cm2 with the same treatment). In addition, ion leakage (from 35.17% in the control to 16.17%) and fruit cracking (from 25.40% in the control to 11.14%) were reduced with the CaCl2 3% + boron 1.5% treatment. Boron plays a role in sugar movement and promotes the formation of fruit buds in plants. In the context of litchi cultivation, the use of borax at a concentration of 0.4% resulted in minimal fruit drop (69.45%), reduced fruit cracking (4.63%), and lighter seed weight (2.30 g). Furthermore, this treatment led to maximum fruit retention (30.55%), fruit set (62.50%), larger fruit dimensions (4.50 cm in length and 3.96 cm in width), greater fruit weight (24.85 g), higher pulp weight (20.73 g), an elevated fruit pulp-to-seed ratio (5.50%), increased fruit yield (120.85 kg per plant), and superior quality characteristics. These quality attributes included higher total soluble solids (22.55°Brix) and increased total sugar content (18.42%) with a lower percentage of titratable acidity (0.41%). These findings were observed in the plains of central Uttar Pradesh, India [54].

Rana and Sharma [55] discovered that grape berries and clusters exhibited an increase in both weight and volume when grapevines were subjected to boron spraying at concentrations of 0.025 and 0.05%. Furthermore, the application of calcium and boron, either individually or in combination, led to larger fruit size and a higher number of fruits per plant in the ber cv. Dongzao, consequently enhancing the fruit yield per plant [56]. In the case of Indian gooseberry and ber, the introduction of boron also contributed to an elevation in vitamin C content [57, 58]. This improvement in fruit quality may be attributed to the role of boron in facilitating carbohydrate transport within plants. Dhaker et al. [59] reported that the use of a foliar spray containing 0.6% borax significantly increased fruit weight (962.0 g) and fruit yield (21.21 kg) in bael (*Aegle marmelos* Corr.). Additionally, this concentration resulted in minimal fruit cracking (2.14%) and reduced peel thickness (2.41 mm), while also enhancing various qualitative characteristics of the fruit. Furthermore, the combined effect of organic manure (50 kg per tree) and foliar application of 0.6% borax significantly increased tree height (49.20 cm), stem girth (2.75 cm), fruit weight (980.0 g), and fruit yield (36.34 q/ha) compared to control plants.

Sotomayor et al. [60] reported that kiwifruits derived from shoots with boron-treated leaves exhibited a 14.1% increase in weight compared to the control, while fruits from boron-treated flowers were 17% heavier than those from untreated flowers. Additionally, significant differences in fruit length were observed between treated and control plants. In the case of papaya cv. Shahi, foliar application of 1.0 kg/ha of boron resulted in the highest fruit yield (49.01 t/ha) [61]. Boron also had a positive impact on pineapple fruit quality [62], where its application at a concentration of 2.0 mg/kg of B was deemed beneficial for enhancing fruit weight, total soluble solids (TSS), the TSS-to-acidity ratio, vitamin C content, and the concentration of aroma volatiles. As a result, the use of boron fertilizer is recommended for pineapple cultivation.

Mangoes exhibited leaf boron (B) deficiency levels ranging from 20 to 49 ppm, with sufficiency levels falling within the 50 to 100 ppm range [63]. Under the agro-climatic conditions of Doon Valley, Uttarakhand, India, an emerging physiological anomaly known as internal necrosis affects developing mango fruits due to boron deficiency, often resulting in fruit cracking. Notably, 'Dashehari' proved to be highly sensitive to both disorders, while 'Chausa' displayed the greatest tolerance. The analysis of nutrients in the leaves and fruit-bearing branches revealed that internal necrosis was primarily caused by boron deficiency. 'Dashehari' exhibited elevated leaf nitrogen levels, which may have contributed to the low levels of leaf boron and subsequently led to the occurrence of internal necrosis disorder. The most effective solution was found to be foliar sprays of boron (in the form of disodium octaborate tetrahydrate) at a concentration of 0.10%, resulting in substantial increases in boron levels in both leaves and fruits (149.64 and 120.14% increases, respectively) of cv. 'Dashehari'. Furthermore, it was observed that the internal necrosis disorder exacerbated fruit cracking in 'Dashehari'. The study also highlighted that foliar application was more effective than soil application in terms of increasing yield while reducing internal necrosis and fruit cracking disorders [64]. In mango cv. SB Chausa, a combined application of KNO_3 (1.0%) and boric acid (0.2%) led to enhanced fruit set (38–42%), increased fruit retention (56–88%), higher fruit weight, and improved yield per plant [65]. Application of Agricol at a rate of 5 grams on two sides of the plant canopy, specifically on the N-E and S-W aspects, resulted in the highest fruit yield, with each tree producing 46.2 and 45.92 kilograms of fruit [66]. Additionally, when Disodium Octaborate Tetrahydrate (DOT) was applied at a rate of 5 grams, it led to the highest number of hermaphrodite flowers per panicle (208.55 and 207.71), the best sex ratio per panicle (0.66 and 0.61), and the lowest fruit drop (51.23 and 50.50). On the other hand, the application of Agricol at 5 grams significantly improved fruit quality parameters and pulp weight (153.23 and 152.04). Maximum fruit dimensions (9.38 and 9.17 centimeters in length and width) were achieved with the use of Chemibor-P, while the application of DOT at 10 grams on the north-east and south-west canopy of the plant resulted in the highest levels of fruit bioactive compounds such as vitamin C (65.31 mg/100 g pulp and 65.64 mg/100 g pulp) and beta-carotene (3330.55 μg/100 g and 3315.18 μg/100 g). When boron was applied to mango cv. Amrapali through soil from various mineral sources, it had a clearly positive impact on the promotion of reproductive growth compared to control plants. The application of boron at two stages, namely pre-flowering and the pea stage of fruit development, influenced mango flowering and fruiting by increasing the number of flowers, reducing fruit drop, and ultimately resulting in higher mango yields. These findings underscore the role of boron in the reproductive physiology of plants, including processes like pollen tube elongation, pollen germination, sugar transport, and carbohydrate synthesis [67].

7. Boron uptake and management

Plant growth and development depends on several factors including nutrient uptake capacity and distribution to other growing parts of the plant [68]. In a single growing season crop plants may face either deficiency or toxicity [69]. This is due to very narrow range of B deficiency and toxicity in soils and plants [70] therefore, it is important to apply optimum B fertilizer for supply of B in deficient soils for normal growth, yield and quality of produce. In gladiolus, B at 0.3% showed marginal increase in flower duration while greater concentration proved toxic and lesser concentration was found to be deficient [71]. The frequency of B application would depend on doses and nature of the crop. The best response was obtained with basal application of B on crops. When boron deficiency arises, it can be addressed by applying foliar sprays containing 2.0–2.5 grams per liter of either boric acid or solubor [15]. To ensure the soil solution maintains optimal boron concentrations for maximum production, it is essential to employ environmentally friendly and advantageous techniques. These approaches not only boost boron absorption and its distribution to various plant components but also enhance soil fertility and crop yield. Some important approaches to enhance the B acquisition are as follows:

7.1 Grafting

Soils experiencing boron deficiency can be addressed by either incorporating boron directly into the soil or by applying boron-containing fertilizers through foliar methods. However, this approach can raise the overall cost of crop cultivation and potentially lead to boron toxicity issues due to the narrow margin between deficiency and toxicity. Therefore, an environmentally friendly and suitable alternative is the utilization of specific rootstocks tailored for different crops [72, 73]. These rootstocks possess the capability to efficiently absorb substantial amounts of boron from the soil and transport it to the upper parts of the plant, ensuring proper physiological functioning. Extensive research demonstrates that various rootstocks significantly improve the nutritional status of plants across different crops, primarily due to their effective water and mineral absorption abilities from the soil solution, surpassing those of self-rooted plants [74–77]. Furthermore, rootstocks enhance the resilience of scion cultivars to both boron deficiency [78] and toxicity [79]. The intricate physiological interactions between the scion and rootstock and their impact on mineral acquisition have been thoroughly explored in various plant species. For instance, in the case of citrus trees, Carrizo citrange (*Citrus sinensis* Osb. × *Poncirus trifoliata* [L.] Raf.) and red tangerine (*C. tangerina*) have been identified as highly effective rootstocks with strong genetic traits for boron uptake and transportation to the upper canopy under conditions of limited boron availability. Trifoliate orange (*P. trifoliata* [L.] Raf.) exhibits moderate efficiency in this regard, while sour orange (*C. aurantium* L.) and fragrant citrus (*C. medica*) have been deemed inefficient rootstocks for boron uptake and transport [72]. In studies by Liu et al. [80, 81], the impact of boron on Carrizo citrange (*C. sinensis* Osb. × *P. trifoliata* [L.] Raf.) and trifoliate orange (*P. trifoliata* [L.] Raf.) rootstocks grafted onto orange plants was examined. Their findings indicated a notable increase in boron absorption and newly absorbed boron concentration in the lower and upper leaves of Carrizo citrange grafted plants when compared to those grafted onto trifoliate orange rootstocks. In the case of pistachio trees (*Pistacia vera* cv. Kerman), *P. atlantica* rootstock demonstrated a high capacity for boron absorption and uptake of other nutrients from the soil solution, resulting in significantly higher concentrations (1.2–2.4 times more) in the leaves [82].

The distribution of ions in various plant parts can significantly differ based on their availability in the soil solution. El-Motaium et al. [79] noted a strong connection between the rootstock and boron uptake in pear plants, resulting in an increase in boron concentration of up to 50–80% in leaves and 100–300% in stems. However, the use of Prunus rootstock did not yield a notable increase in boron content in the root tissues. In the case of Newhall orange (*C. sinensis* Osb.), Sheng et al. [83] observed that when exposed to limited boron supply, boron content decreased in leaf (23–53%) and scion (40–65%) tissues but increased in rootstock parts (35–60%) when grafted onto Carrizo citrange (*C. sinensis* Osb. × *P. trifoliata* [L.] Raf.) as compared to trifoliate orange (*P. trifoliata* [L.] Raf.). Wang et al. [84] conducted a study on boron absorption patterns in four citrus rootstock-scion combinations and found that, under inadequate boron supply, the maximum boron concentration was observed in the buds and leaves of *C. sinensis* [L.] Osb. cv. Fengjie-72 when grafted onto Carrizo citrange and trifoliate orange plants. Notably, the boron accumulation in the Newhall scion grafted onto Carrizo citrange was higher (24%) compared to other combinations. Furthermore, a higher proportion of available boron (36%) was detected in the leaves of Carrizo citrange when compared to plants grafted onto trifoliate orange.

7.2 Biostimulators

Biostimulants are substances, distinct from fertilizers, soil conditioners, or pesticides, that have the capacity to impact various metabolic processes in plants, such as cell division, respiration, photosynthesis, and ion absorption, even when applied in small quantities [85]. These biostimulators can be employed to augment mineral uptake in plants while requiring minimal inputs. In recent times, biostimulants have played a pivotal role in altering plant physiology and optimizing plant growth [86]. They engage with the plant's signaling pathways to mitigate adverse plant responses during stressful conditions, ultimately promoting optimal plant development. Crops treated with biostimulants exhibit greater resilience to challenging environmental circumstances and demonstrate enhanced efficiency in ion absorption when faced with limited ion availability, primarily due to improved antioxidant production [87]. Humic substances (HS), a type of organic biostimulant, are renowned for their ability to enhance soil structure and root architecture by increasing the activity of root H + -ATPase. Consequently, they find widespread application in ion acquisition, with the effectiveness depending on factors such as concentration, plant species, and environmental conditions [88]. Field trials involving the application of biostimulants to the soil of *Vicia faba* cv. Giza beans demonstrated enhanced soil structure and ion uptake compared to untreated controls [89]. Conversely, the use of composted sewage sludge containing HS resulted in improved growth and yield of *Capsicum annuum* L. cv. Piquillo [90]. Additionally, these benefits were associated with increased availability of micronutrients in the substrate and enhanced microbial activity within plants. This microbial activity aids in reducing ion leaching by lowering soil pH through the production of organic acids like citrate, oxalate, and malate. HS forms complexes with micronutrients, and the plant's plasma membrane generates a proton motive force that facilitates the active and passive transport of ions through the symplastic pathway, thereby increasing the availability of trace elements to plants [91].

7.3 Mycorrhizal fungi (MF)

In rough lemon (*Citrus jambhiri* Lush), the application of boron through foliar spraying and soil amendment, coupled with Glomus fasciculatum inoculation, not only led to an increase in total boron accumulation in the leaves by 11–18% but also resulted in enhanced exudation of root sugars and amino acids when compared to plants that were not inoculated [92]. The presence of arbuscular mycorrhizal fungi (AMF) in the soil can impact boron concentration in plants. Research findings on the effect of AMF inoculation vary, with some studies reporting reduced boron acquisition in shoots of MF-inoculated plants [93], others showing no significant effect [94], and yet others indicating enhanced boron acquisition [95]. However, the precise role of mycorrhizal activity in relation to boron remains unproven and necessitates further investigation. The vascular structure found in higher plants underlines the importance of boron in lignification [96]. While passive uptake of boric acid seems to occur in plants, the mechanism behind mycorrhizal boron uptake is not yet fully understood. The critical role of sugar alcohols such as sucrose, sorbitol, and mannitol in the remobilization of boron within plant tissues is well-documented [4, 97]. According to Lewis's hypothesis [96], sucrose, due to its low affinity for boron in vascular plants, is primarily responsible for boron mobilization. In contrast, fungal carbohydrates, especially mannitol, exhibit a high affinity for forming complexes with boron, leading to limited boron mobility from the fungal symbiotic partner to the host plant. However, certain mycelia have been observed to facilitate the mobility of the mannitol-boron complex, allowing for the continuous uptake and long-distance transport of boron in plants [6]. While it is established that the application of AMF can aid in boron acquisition and transport within plants, there are still specific aspects of this process that require further investigation.

7.4 Plant-growth-promoting rhizobacteria (PGPR)

Rhizobacteria, also referred to as plant-growth-promoting rhizobacteria (PGPR), are beneficial and actively root-colonizing microorganisms that establish a symbiotic relationship with plant roots. They play crucial roles in various agricultural aspects, including nitrogen fixation [98], improving tolerance to salinity and drought [99], producing enzymes that combat pathogenic microorganisms, solubilizing nutrients, and generating phytohormones like IAA, cytokinins, and gibberellins, which stimulate root growth [100]. This root proliferation, in turn, enhances water and nutrient uptake by plants.

In the case of lentils (*Lens culinaris* Medik), the inoculation of PGPR not only increased nitrogen (N) uptake (2.26–2.95%) and phosphorus (P) uptake (0.52–0.82%) in the roots, stems, and grains but also improved plant growth parameters such as root and shoot length, as well as their fresh and dry weights. Additionally, the application of PGPR resulted in higher levels of phytohormones (IAA, GA_3) and increased macro- and micronutrient concentrations in crops like *Raphanus sativus* and Musa spp. [101, 102]. In potatoes (*Solanum tuberosum* L.), the P content saw a 43.1% improvement with the introduction of the *Bacillus cereus* P31 strain, while the *Achromobacter xylosoxidans* strain P35 increased N and K content by up to 50.5 and 48.3%, respectively [103].

Numerous studies support the role of bacteria in absorbing excessive boron levels from soil solutions. These studies involve B-tolerant bacterial strains belonging to genera such as *Bacillus, Gracilibacillus, Lysinibacillus, Boronitolerans, Variovorax, Pseudomonas, Mycobacterium*, and *Rhodococcus*, which are capable of absorbing toxic levels of boron from the soil [104, 105]. Cheke et al. [106] reported that the availability of micronutrients (Fe, Mn, Cu, Zn, and B) was higher in the rhizosphere soil compared to non-rhizosphere soil, indicating that the tree rhizosphere has an impact on the availability of trace elements in the soil. In a study conducted in Nagaland, India, rhizosphere soils collected from healthy Khasi mandarin (*Citrus reticulata* Blanco) plants in an orchard displayed a higher bacterial population compared to the fungal population in the rhizosphere of four high-yielding plants [107]. However, additional research is necessary to identify efficient boron-capturing bacteria that can enhance boron availability for crops under conditions of limited boron supply.

7.5 Nanotechnology

Nanotechnology presents an innovative strategy applicable in agriculture for managing both biotic and abiotic stress, detecting diseases, and improving nutrient absorption [108, 109]. This cutting-edge technology is essential to address the challenges of limited nutrient and water resources while aiming to increase the production of high-quality agricultural crops. Nanotechnology enhances plant production and nutrient utilization efficiency by requiring fewer resources compared to traditional methods. Nanoparticles (NPs) have unique physicochemical properties that positively impact plant metabolism, leading to increased crop yield and nutritional value [110]. To illustrate, the use of copper NPs in watermelon cultivation resulted in improved plant growth and development compared to the control group [111]. Baruah and Dutta [112] demonstrated that hydrogels and zeolites have the capacity to absorb environmental contaminants and enhance soil water retention. Chitosan NPs have proven effective in reducing fertilizer consumption, contributing to a reduction in environmental pollution [108]. While the application of CeO_2 and ZnO NPs did not lead to an increase in macronutrient concentrations in *Cucumis sativus* fruit, they did elevate micronutrient levels [113]. The use of nano-titanium dioxide (TiO_2) boosted chlorophyll synthesis and photosynthetic activity by enhancing ion uptake efficiency in spinach [114]. Consequently, adopting nanotechnology approaches is a promising strategy for enhancing boron uptake and utilization efficiency in plants, potentially reducing the need for boron fertilizer in crop cultivation.

8. Conclusions

Boron (B) is a vital trace element essential for the proper physiological functioning of higher plants. B deficiency represents a nutritional disorder that has detrimental effects on plant metabolism and growth. The ability of crops to efficiently utilize B resources can vary significantly, and it is crop-specific. Thus, from an agricultural perspective, there is a necessity to identify key cultivars among agronomic and horticultural crops, as well as different conditions that enable optimal utilization of available B resources, especially those that thrive under B-deficient conditions. The intricate relationship between rootstock and scion requires further in-depth studies to identify exceptional root systems, particularly those indigenous to specific regions that exhibit tolerance to both B deficiency and toxicity. Grafting and arbuscular mycorrhizal fungi (AMF) inoculation have been shown to enhance various aspects of plant physiology and nutrition, with several studies highlighting their crucial role in B uptake. Additionally, there is potential for molecular-level investigations into the role of B in plants, which could pave the way for novel strategies to enhance B stress tolerance in crops. Nanotechnology is an emerging agricultural technique designed to address plant nutrition-related challenges. The combination of these techniques has the potential to improve B uptake. Research has demonstrated that the simultaneous use of grafting and copper nanoparticles (NPs) can enhance the growth and development of watermelon by increasing ion uptake. In certain plant species, the combined inoculation of mycorrhizal fungi (MF) and plant-growth-promoting rhizobacteria (PGPR) has improved growth by augmenting water and macronutrient levels. Therefore, these existing techniques should be harnessed and further refined, considering crop-specific and location-specific factors, to achieve better outcomes and enhance B uptake and utilization in plants.

Author details

Pauline Alila
Department of Horticulture, School of Agricultural Sciences, Nagaland University, Medziphema Campus, Nagaland, India

*Address all correspondence to: paulinealila@gmail.com

References

[1] Warington K. The effect of boric acid and borax on broad bean and certain other plants. Annals of Botany. 1923;**37**:629-672

[2] Marschner H. Mineral Nutrition of Higher Plants. 2nd ed. London: Academic Press; 1995. 889 p

[3] Oertli JJ. The mobility of boron in plants. Plant and Soil. 1993;**155**(156):301-304

[4] Brown PH, Hu H. Phloem mobility of boron is species dependent. Evidence for phloem mobility in sorbitol rich species. Annals of Botany. 1996;**77**:497-505

[5] Brown PH, Shelp BJ. Boron mobility in plants. Plant and Soil. 1997;**193**:85-101

[6] Hu H, Brown PH. Absorption of boron by plant roots. Plant and Soil. 1997;**193**:49-58

[7] Stangoulis J, Tate M, Graham R, Bucknall M, Palmer L, Boughton B, et al. The mechanism of boron mobility in wheat and canola phloem. Plant Physiology. 2010;**153**:876-881

[8] Nyomora AMS, Brown PH, Pinney K, Polito VS. Foliar application of boron to almond trees affects pollen quality. Journal of American Society of Horticultural Science. 2000;**125**:265-270

[9] Thompson M. Flowering, pollination and fruit set. In: Webster AD, Looney NE, editors. Cherries: Crop Physiology, Production and Uses. Wallingford: CAB International; 1996. pp. 223-242

[10] Dregne HE, Power WL. Boron fertilization of alfalfa and other legumes in Oregon. Journal of American Society of Agronomy. 1942;**34**:902-912

[11] Wojcik P. 'Schattenmorelle' tart cherry response to boron fertilization. Journal of Plant Nutrition. 2006;**29**(9):1709-1718

[12] Gupta UC. Introduction. In: Gupta UC, editor. Boron and its Role in Crop Production. Boca Raton, FL, U.S.A: CRC Press; 1993. p. 237

[13] Rerkasem B, Jamjod S. Genotypic variation in plant response to low boron and implications for plant breeding. Plant and Soil. 1997;**193**:169-180

[14] Shorrocks VM. The occurrence and correction of boron deficiency. Plant and Soil. 1997;**193**:121-148

[15] Prasad R, Kumar D, Shivay YS, Rana DS. Boron in Indian agriculture-a review. Indian Journal of Agronomy. 2014;**59**(4):511-517

[16] Behera SK, Singh MV, Lakaria BL. Micronutrient deficiencies in Indian soils and their amelioration through fertilization. Indian Farming. 2009;**5**(2):28-31

[17] Abid N, Ibrahim M, Ahmad N, Anwar SA. Boron contents of light and medium textured soils and cotton plants. International Journal of Agriculture and Biology. 2002;**4**(4):534-536

[18] Sakal R, Singh AP. Boron research and agricultural production. In: Tandon HLS, editor. Micronutrient Research and Agricultural Production. New Delhi, India: Fertilizer Development and Consultation Organization; 1995. pp. 1-64. ISBN: 8185116601

[19] Tiwari KN, Tiwari R. Managing boron deficiency in vegetable crops. Indian Horticulture. 2015;**60**(1):16-19

[20] Belvins DG, Lukaszewski KM. Boron in plant structure and function. Annual Review of Plant Physiology and Plant Molecular Biology. 1998;**49**:481-500

[21] Das SK, Avasthe RK, Yadav A. Secondary and micronutrients: Deficiency symptoms and management in organic farming. Innovative Farming. 2017;**2**(4):209-211

[22] Singh AK, Singh AK, Singh JK. Boron in crop production from soil to plant system: A review. Archives of Agriculture and Environmental Science. 2020;**5**(2):218-222

[23] Shekhawat K, Shivay YS. Effect of nitrogen sources, Sulphur and boron levels on productivity, nutrient uptake and quality of sunflower (Helianthus annuus). Indian Journal of Agronomy. 2008;**53**(2):129-134

[24] Sinha RB, Sakal R, Singh AP, Bhogal NS. Response of some field crops to boron application in calcareous soils. Journal of the Indian Society of Soil Science. 1991;**39**(1):118-122

[25] Mani PK, Haldar M. Effect of dolomite on boron transformation in acid soil in relation to nutrition green gram. Journal of the Indian Society of Soil Science. 1996;**44**(3):458-461

[26] Ansari MA, Prakash N, Singh IM, Sharma PK, Punitha P. Efficacy of boron sources on productivity, profitability and energy use efficiency of groundnut (Arachis hypogaea) under north east hill regions. Indian Journal of Agricultural Sciences. 2013;**83**(9):959-963

[27] Ateeque M, Malewar GU, More SD. Influence of phosphorus and boron on yield and chemical composition of sunflower. Journal of the Indian Society of Soil Science. 1993;**41**(1):100-102

[28] Naik KGA, Manure GR, Badiger MK. Yield of castor by fertilizing with phosphorus, Sulphur and boron. Journal of the Indian Society of Soil Science. 1993;**41**(4):686-688

[29] Dwivedi GK, Dwivedi M, Pal SS. Mode of application of micronutrients in soybean-wheat crop sequence. Journal of the Indian Society of Soil Science. 1990;**38**:458-463

[30] Mondal C, Bandopadhyay P, Alipatra A, Banerjee H. Performance of summer mungbean [*Vigna radiata* (L.) Wilczek] under different irrigation regimes and boron levels. Journal of Food Legumes. 2012;**25**(1):37-40

[31] Sarkar S, Banerjee H, Ray K, Ghosh D. Boron fertilization effects in processing grade potato on an Inceptisol of West Bengal, India. Journal of Plant Nutrition. 2018;**41**(11):1456-1470

[32] Vitosh ML, Warncke DD, Lucal RE. Secondary and Micronutrients for Vegetables and Field Crops. Michigan State University, Fram Science Series, MSU Extension Publication Archive; 1994. pp. 10-12

[33] Combrink NJJ, Maree PCJ. The effect of calcium and boron on the quality of muskmelon (*Cucumis melo* L.). Journal of the South African Society for Horticultural Science. 1995;**5**:33-38

[34] Kotur SC, Kumar S. Response of cauliflower (Brassica oleracea var. botrytis) to boron in Chotanagpur region. Indian Journal of Agricultural Sciences. 1989;**59**(10):640-644

[35] Haque ME, Paul AK, J. R. Sarker effect of nitrogen and boron on the growth and yield of tomato (*Lycopersicon esculentum* M.). International Journal of Bio-Resource and Stress Management. 2011;**2**(3):277-282

[36] Naz RMM, Muhammad S, Hamid A, Bibi F. Effect of boron on the flowering and fruiting of tomato. Sarhad Journal of Agriculture. 2012;**28**(1):37-40

[37] Ganeshamurthy AN. Effect of directly-applied and residual boron on nutrition in French bean-cabbage cropping sequence under Alfisol. Journal of Horticultural Sciences. 2014;**9**(2):185-190

[38] Solanki VPS, Singh J, Singh V. Differential response of vegetable crops to boron application. Annals of Plant and Soil Research. 2018;**20**(3):239-242

[39] Loomis WD, Durst RW. Chemistry and biology of boron. BioFactors. 1992;**3**:229-239

[40] Peryea FJ. Boron nutrition in deciduous tree fruits. In: Peterson AB, Stevens RG, editors. Tree Fruit Nutrition. Yakima, Washington: Good Fruit Grower; 1994. pp. 95-99

[41] Shear CB, Faust M. Nutritional ranges in deciduous tree fruits and nuts. Horticulture Review. 1980;**2**:142-163

[42] Wojcik P, Wojcik M. Effect of boron fertilization on sweet cherry tree yield and fruit quality. Journal of Plant Nutrition. 2006;**29**:1755-1766

[43] Ruchal OK, Pandeya SR, Regmia R, Regmib R, Bahadur B. Effect of foliar application of micronutrient (zinc and boron) in flowering and fruit setting of mandarin (*citrus reticulata* Blanco) in Dailekh, Nepal. Magratic Malaysian Journal of Sustainable Agriculture (MJSA). 2020;**4**(2):94-98

[44] Sheikh KHA, Singh B, Haokip SW, Shankar K, Debbarma R, Gaitri Devi A, et al. Response of yield and fruit quality to foliar application of micronutrients in lemon [Citrus Limon (L.) Burm.] cv. Assam lemon. Journal of Horticultural Sciences. 2021;**16**(2):144-153

[45] Arora JS, Singh JR. Responses of guava (Psidium guajava L.) to boron spray. Journal of Japanese Society of Horticultural Sciences. 1972;**41**(3):239-244

[46] Rajput CBS, Chand S. Significance of boron and zinc in guava (Psidium guajava L.). Bangladesh Horticulture. 1975;**3**(2):22-27

[47] Singh RR, Joon MS, Daulta BS. A note on the effect of foliar spray of urea and boric acid on physico-chemical composition of guava fruits cv. L-49. Haryana Journal of Horticultural Sciences. 1983;**12**(1-2):68-72

[48] Pandey DK, Pathak RA, Pathak RK. Studies in foliar application of nutrients and plant growth regulators in Sardar guava (Psidium guajava L.). Indian Journal of Horticulture. 1988;**45**(3-4):197-102

[49] Ali W, Pathak RA, Yadav AL. Effect of foliar application of nutrients on guava (Psidium guajava L.) cv. Allahabad Safeda. Progressive Horticulture. 1991;**23**(1-4):18-21

[50] Yadav PK. Note on yield and quality parameters of guava as influenced by foliar application of nutrient and plant growth regulators. Current Agriculture. 1998;**22**(1-2):117-119

[51] Bhatia SK, Yadav S, Ahlawat VP, Dahiya SS. Effect of foliar application of nutrients on the yield and fruit quality of winter season guava cv. L-49. Haryana Journal of Horticultural Sciences. 2001;**30**(1-2):6-7

[52] Upreti A, Kumar G. Effect of mineral nutrient sprays on yield and quality

of litchi fruits (*Litchi chinensis* Sonn.) cv. Rose scented. Indian Journal of Horticulture. 1996;**53**(2):121-124

[53] Haq I, Rab A. Foliar application of calcium chloride and borax affects the fruit skin strength and cracking incidence in litchi (*Litchi chinensis* Sonn.) cultivars. African Journal of Biotechnology. 2012;**11**(10):2445-2453

[54] Gupta A, Tripathi VK, Shukla JK. Influence of GA_3, zinc and boron on fruit drop, yield and quality of litchi (Litchi chinensis Sonn). Biological Forum – An International Journal. 2022;**14**(3):1079-1083

[55] Rana RS, Sharma HC. Effect of boron on yield and quality of grapes (*Vitis vinifera*). Indian Journal of Horticulture. 1979;**36**(3):275-277

[56] Mao YM, Shen LY, Wei W, Wang XL, Hu YL, Xu SS, et al. Effects of foliar applications of boron and calcium on the fruit quality of' Dongzao (*Zizyphus jujuba* mill.). In: Proceedings of the International Horticultural Congress on Horticulture: Sustaining Lives, Livelihoods and Landscapes. Istanbul: ASHS Editions; 2014. pp. 105-108

[57] Shukla K. Indian Journal of Agricultural Sciences. 2011;**81**:628-632

[58] Samant D, Mishra NK, Singh AK, Lal RL. Indian Journal of Horticulture. 2008;**65**:399-404

[59] Dhaker M, Soni AK, Yadav PK, Chandra A, Kumar H. Response of different levels of farm yard manure and boron on growth and yield of bael (Aegle marmelos Corr.). Asian Journal of Horticulture. 2013;**8**(2):767-771

[60] Sotomayor C, Norambuena P, Ruiz R. Boron dynamics related to fruit growth and seed production in kiwifruit (Actinidia deliciosa cv. Hayward). Ciencia e investigación agraria: revista latinoamericana de ciencias de la agricultura. 2010;**37**(1):133-141

[61] Mollah M, Rahim M, Islam M, Khatun M, Rahman M. Response of papaya varieties to basal and foliar application of boron. International Journal of Sustainable Agricultural Technology. 2006;**2**(1):28-31

[62] Wei C, Ma Z, Liu Y, Qiao J, Sun G. Effect of boron on fruit quality in pineapple. In: 2017 International Conference on Biotechnology and Bioengineering (ICBB-2017). AIP Conference Proceedings 1956. Offenburg, Germany: AIP Publishing; 2018. pp. 020006-1-020006-6. DOI: 10.1063/1.5034258

[63] Singh VK. CISH Lucknow. Data from Farming outlook. 2007;**5**:3

[64] Saran PL, Kumar R. Boron deficiency disorders in mango (Mangifera indica): Field screening, nutrient composition and amelioration by boron application. Indian Journal of Agricultural Sciences. 2011;**81**(6):506-510

[65] Bibi F, Ahmad I, Bakhsh A, Kiran S, Danish S, Ullah H, et al. Effect of foliar application of boron with calcium and potassium on quality and yield of mango cv. Summer Bahisht (SB) Chaunsa. Open Agriculture. 2019;**4**:98-106

[66] Pavithra G, Alila P, Maiti CS, Sarkar A, Sahu AK. Effect of mineral boron sources on reproductive, yield and quality characteristics of mango (Mangifera indica) current. Horticulture. 2021;**9**(2):41-45

[67] Pavithra G, Alila P, Maiti CS, Sarkar A, Sahu AK. Effect of different sources of boron on flowering and yield

of mango cv. Amrapali. The Pharma Innovation Journal. 2022;**11**(4):909-913

[68] Brown PH, Bellaloui N, Wimmer MA, Bassil ES, Ruiz J, Hu H, et al. Boron in plant biology. Plant Biology. 2002;**4**:205-223

[69] Batabyal K, Sarkar D, Mandal B. Critical levels of boron in soils for cauliflower (Brassica oleracea var. botrytis). Journal of Plant Nutrition. 2015;**38**:1822-1835

[70] Singh MV, Goswami V. Efficiency of boron fortified NPK fertilizer in correcting boron deficiency in some cereal and oilseeds crops in India. In: Proceedings Book, 2013 17th International Plant Nutrient Colloquium. Istanbul, Turkey. 17-18 August 2013

[71] Somkumar AR, Singh AK, Sisidia A, Lamsal A, Giri S. Effect of boron on growth and flowering in gladiolus (*gladiolus* sp.). Current Horticulture. 2023;**11**(2):56-59

[72] Mei L, Sheng O, Peng SA, Zhou GF, Wei QJ, Li QH. Growth, root morphology and boron uptake by citrus rootstock seedlings differing in boron-deficiency responses. Scientia Horticulturae. 2011;**129**:426-432

[73] Zhou FG, Peng SA, Liu YZ, Wei QJ, Han J, Islam MZ. The physiological and nutritional responses of seven different citrus rootstock seedlings to boron deficiency. Trees. 2014;**28**:295-307

[74] Ahmed W, Nawaz MA, Iqbal MA, Khan MM. Effect of different rootstocks on plant nutrient status and yield in Kinnow mandarin (*Citrus reticulata* Blanco). Pakistan Journal of Botany. 2007;**39**:1779-1786

[75] Huang Y, Jiao Y, Nawaz MA, Chen C, Liu L, Kong Q, et al. Improving magnesium uptake, photosynthesis and antioxidant enzyme activities of watermelon by grafting onto pumpkin rootstock under low magnesium. Plant and Soil. 2016;**409**:229-246

[76] Nawaz MA, Imtiaz M, Kong Q, Cheng F, Ahmad W, Huang Y, et al. Grafting: A technique to modify ion accumulation in horticultural crops. Frontiers in Plant Science. 2016;7: 1457-1457. DOI: 10.3389/fpls 2016.01457

[77] Nawaz MA, Wang L, Jiao Y, Chen C, Zhao L, Mei M, et al. Pumpkin rootstock improves nitrogen use efficiency of watermelon scion by enhancing nutrient uptake, cytokinin content, and expression of nitrate reductase genes. Plant Growth Regulation. 2017;**82**:233-246

[78] Wojicik P, Wojcik M, Treder W. Boron absorption and translocation in apple rootstock under conditions of low medium boron. Journal of Plant Nutrition. 2003;**26**:961-968

[79] El-Motaium R, Hu H, Brown PH. The relation tolerance of six prunus rootstocks to boron and salinity. Journal of the American Society for Horticultural Science. 1994;**119**:1169-1175

[80] Liu G, Jiang C, Wang Y, Peng SA, Zhong B, Ceng Q, et al. Changes in mineral element contents of Newhall' navel orange (*Citrus sinensis* Osb.) grafted on two different rootstocks under boron deficiency. Plant Nutrition and Fertilizer Science. 2011;**1**:180-185

[81] Liu GD, Wang RD, Wu LS, Peng SA, Wang YH, Jiang CC. Boron distribution and mobility in navel orange grafted on citrange and trifoliate orange. Plant and Soil. 2012;**2012**(360):123-133

[82] Brown PH, Zhang Q, Ferguson L. Influence of rootstock on nutrient

acquisition by pistachio. Journal of Plant Nutrition. 1994;**17**:1137-1148

[83] Sheng O, Song S, Peng S, Deng X. The effects of low boron on growth, gas exchange, boron concentration and distribution of 'Newhall' navel orange (*Citrus sinensis* Osb.) plants grafted on two rootstocks. Scientia Horticulturae. 2009;**121**:278-283

[84] Wang N, Yan T, Fu L, Zhou G, Liu YZ, Peng SA. Differences in boron distribution and forms in four citrus scion–rootstock combinations with contrasting boron efficiency under boron-deficient conditions. Trees. 2014;**28**:1589-1598

[85] Kuffaman GL, Kneivel DP, Watschke TL. Effect of a biostimulants on the heat tolerance associated with photosynthetic capacity, membrane stability and polyphenol production of perennial rye grass. Crop Science. 2007;**47**:261-267

[86] Brown P, Saa S. Biostimulants in agriculture. Frontiers in Plant Science. 2015;**6**:671

[87] Sharma HSS, Fleming C, Selby C, Rao JR, Martin T. Plant biostimulants: A review on the processing of macro algae and use of extracts for crop management to reduce abiotic stresses. Journal of Applied Phycology. 2014;**26**:465-490

[88] Halpern M, Bar-Tal A, Ofek M, Minz D, Muller T, Yermiyahu U. The use of biostimulants for enhancing nutrient uptake. In: Sparks DL, editor. Advances in Agronomy. 1st ed. Vol. 147. New York, NY, USA: Elsevier Inc; 2015. pp. 141-174

[89] Abbas SM. The influence of biostimulants on the growth and on the biochemical composition of *Vicia faba* CV. Giza 3 beans. Romanian Biotechnology Letters. 2013;**18**:8061-8068

[90] Pascual I, Azcona I, Aguirreolea J, Morales F, Corpas FJ, Palma JM, et al. Growth, yield and fruit quality of pepper plants amended with two sanitized sewage sludges. Journal of Agricultural and Food Chemistry. 2010;**58**:6951-6959

[91] Canellas LP, Olivares FL, Aguiar NO, Jones DL, Nebbioso A, Mazzei P, et al. Humic and fulvic acid as biostimulants in horticulture. Scientia Horticulturae. 2015;**196**:15-27

[92] Dixon RK, Garrett HE, Cox GS. Boron fertilization, vesicular-arbuscular mycorrhizal colonization and growth of *Citrus jambhiri* lush. Journal of Plant Nutrition. 1989;**1989**(12):687-700

[93] Clark RB, Zobel RW, Zeto SK. Effects of mycorrhizal fungus isolate on mineral acquisition by *Panicum virgatum* in acidic soil. Mycorrhiza. 1999;**9**:167-176

[94] Lu S, Miller HM. The role of VA mycorrhizae in the absorption of P and Zn by maize in field and growth chamber experiments. Canadian Journal of Soil Science. 1989;**69**:97-109

[95] Kothari SK, Marschner H, Römheld V. Direct and indirect effects of VA mycorrhizal fungi and rhizosphere microorganisms on acquisition of mineral nutrients by maize (*Zea mays* L.) in a calcareous soil. The New Phytologist. 1990;**116**:637-645

[96] Lewis DH. Boron, lignification and the origin of vascular plants-a unified hypothesis. The New Phytologist. 1980;**84**:209-229

[97] Liakopoulos G, Stavrianakou S, Filippou M, Fasseas C, Tsadilas C, Drossopoulos I, et al. Boron remobilization at low boron supply in

olive (*Olea europaea)* in relation to leaf and phloem mannitol concentrations. Tree Physiology. 2005;**25**:157-165

[98] Shin F, Cakmakci R, Kantar F. Sugar beet and barley yields in relation to inoculation with N2-fixing and phosphate solubilizing bacteria. Plant and Soil. 2004;**265**:123-129

[99] Zahir ZA, Munir A, Asghar HN, Shahroona B, Arshad M. Effectiveness of rhizobacteria containing ACC-deaminase for growth promotion of peas (*Pisum sativum*) under drought conditions. Journal of Microbiology and Biotechnology. 2008;**18**:958-963

[100] Lucy M, Reed E, Glick BR. Application of free living plant growth-promoting rhizobacteria. Anton Leeuw. 2004;**86**:1-25

[101] Yildrim E, Donmez MF, Turan M. Use of bioinoculants in ameliorative effects on radish plants under salinity stress. Journal of Plant Nutrition. 2008;**31**:2059-2074

[102] Kavinoa M, Harishb S, Kumara N, Saravanakumar D, Samiyappan R. Effect of chitinolytic PGPR on growth, yield and physiological attributes of banana (*Musa* spp.) under field conditions. Applied Soil Ecology. 2010;**45**:71-77

[103] Dawwam GE, Elbeltagy-Emara AHM, Abbas IH, Hassan MM. Beneficial effect of plant growth promoting bacteria isolated from the roots of potato plant. Annals of Agricultural Science. 2013;**58**:195-201

[104] Ahmed I, Yokota A, Fujiwara T. A novel highly boron tolerant bacterium, *bacillus boroniphilus* sp. nov., isolated from soil, that requires boron for its growth. Extremophiles. 2007;**11**:217-224

[105] Miwa H, Fujiwara T. Isolation and identification of boron-accumulating bacteria from contaminated soils and active sludge. Soil Science & Plant Nutrition. 2009;**55**:643-646

[106] Cheke AS, Patil VD, Srivastava AK. Studies on rhizosphere hybridization and nutrient dynamics in sweet orange seedling from pot culture experiment. Journal of Pharmacognosy and Phytochemistry. 2018;**SPI**:3077-3082

[107] Dzüvichü M, Alila P, Jamir S, Kanaujia SP, Srivastava AK. Yield response of rhizosphere hybridization in Khasi mandarin. International Journal of Innovative Horticulture. 2023;**12**(1):102-108

[108] Kashyap PL, Xiang X, Heiden P. Chitosan nanoparticle based delivery systems for sustainable agriculture. International Journal of Biological Macromolecules. 2015;**77**:36-51

[109] Wu H, Shabala L, Shabala S, Giraldo JP. Hydroxyl radical scavenging by cerium oxide nanoparticles improves Arabidopsis salinity tolerance by enhancing leaf mesophyll potassium retention. Environmental Science. Nano. 2018;**5**:1567-1583

[110] Ghormade V, Deshpande MV, Paknikar KM. Perspectives for nano-biotechnology enabled protection and nutrition of plants. Biotechnology Advances. 2011;**29**:792-803

[111] Gómez HG, Godina FR, Ortiz HO, Mendoza AB, Torres VR, De-la-Fuente MC. Use of chitosan-PVA hydrogels with copper nanoparticles to improve the growth of grafted watermelon. Molecules. 2017;**22**:1031

[112] Baruah S, Dutta J. Nanotechnology applications in pollution sensing and degradation in agriculture: A review.

Environmental Chemistry Letters. 2009;**7**:191-204

[113] Zhao L, Peralta-Videa JR, Rico CM, Hernandez-Viezcas JA, Sun Y, Niu G, et al. CeO_2 and ZnO nanoparticles change the nutritional qualities of cucumber (*Cucumis sativus*). Journal of Agricultural and Food Chemistry. *2014*;**62**:2752-2759

[114] Zheng L, Hong F, Lu S, Liu C. Effect of nano-TiO_2 on strength of naturally aged seeds and growth of spinach. Biological Trace Element Research. 2005;**104**:84-91

Section 2

Role and Applications of Boron and Its Compounds in Industrial and Nanotechnology Fields

Chapter 4

Boron and Boron Compounds in Radiation Shielding Materials

Ahmet Hakan Yilmaz, Bülend Ortaç and Sevil Savaskan Yilmaz

Abstract

A risk to the nuclear industry is radiation, specifically neutron radiation. In order to maintain a safe workspace for workers, better shielding is being developed. Current shielding methods are examined and boron is considered a potential material for shielding. All living beings and non-living things on earth are exposed to the daily radiation of natural radiation sources in the air, water, soil, and even in their bodies, as well as artificial radiation sources produced by humans. To be safeguarded from the detrimental influences of radiation, it is important to be careful about three basic issues: time, distance, and shielding. The longer the exposure time to radiation from the radioactive source or the closer one is to the radioactive source, the higher the radiation dose to be received. The radiation emitted by some radionuclides is so intense that you can be exposed to it even though you cannot see it from miles away. It can only be protected from the effects of such intense radioactive materials with strong shielding. Boron, having a large cross-section, is combined with other materials in order to obtain the desired material properties to have shielding that can be applied in different situations.

Keywords: boron, boron compounds, shielding materials, radiation, gamma ray, neutron, polymers, glasses

1. Introduction

Radiation is the emission or transmission of energy in the form of waves, particles, or electromagnetic radiation through space or matter. It can be produced by the sun, radioactive elements, X-ray machines, and nuclear reactions, among other things [1–3]. There are two types of radiation: ionizing and non-ionizing radiation. Ionizing radiation is powerful enough to knock electrons off atoms, causing damage to living tissue and DNA. Non-ionizing radiation, such as visible light and radio waves, lacks the energy to cause this type of harm. Radiation is used in various fields, including medicine, industry, and research, but excessive exposure can harm human health.

In daily life, we can encounter high-energy radiation such as alpha, beta particle emissions, X-ray or gamma-ray, or neutron particle emissions in any form, for example, in many various industrial products, including nuclear power plants, in the health sector, both in diagnosis and treatment and in the aviation field [4]. Any of these radiations that we are unintentionally exposed to can be life threatening for us. However, the consequences of such exposures depend on various factors, such as the type of radiation and the energy associated with it, the amount of absorbed dose, exposure time, and so on.

The radiation energies of galactic cosmic rays, solar particle events, medical X-rays, gamma rays, electrons, and neutrons can vary dramatically depending on the source and particle energy. Here are some rough energy ranges for each type of radiation: High-energy particles that originate outside of our solar system are known as galactic cosmic rays. They can range in energy from a few MeV to several hundred TeV, with some rare events exceeding 1020 eV. Solar particle events are bursts of high-energy particles that originate from the sun. These particles have energies ranging from a few MeV to several GeV, with the most energetic events reaching tens of GeV. Medical X-rays are a type of electromagnetic radiation that is used in medical imaging. X-rays used in medical imaging can have energies ranging from a few keV (thousand electron volts) to several MeV. Gamma rays are a type of electromagnetic radiation with extremely high energies. They can be generated by a variety of processes, including nuclear reactions and astronomical phenomena. Gamma-ray energies can range from a few keV to several TeV. Electrons are subatomic particles that have a negative charge. Depending on the source, their energies can range from a few keV to several GeV. Neutrons are subatomic particles that have no charge. Their energies can also vary greatly depending on the source and method of production, ranging from a few MeV to several GeV.

Radiation therapy is a common cancer treatment method. Cancer cells are stopped or killed using high-energy radiation. Radiation therapy can also be used to treat some benign tumors as well as certain blood diseases (e.g., Hodgkin lymphoma). Radiation therapy, also known as radiotherapy, is a type of targeted therapy. While radiation only affects cancer cells, it has little effect on normal cells. As a result, radiation therapy is frequently combined with other cancer treatment modalities (e.g., chemotherapy). Bone diseases can be treated with radiation therapy. Bone cancer (such as osteosarcoma), lymphoma, or multiple myeloma may have spread to the bones or formed tumors in the bones. Radiation therapy is used to shrink or destroy bone tumor formations to treat these types of cancer. Regional pain and bone fractures can also be treated with radiation therapy. Bone metastases (spreading cancer from another site to the bones) frequently cause pain and fractures.

Neutrons are neutral (zero-charged) particles that are used in nuclear power plants. They can easily pass through most materials and interact with the target atom's nucleus. The majority of sources that emit X-rays and rays also emit neutrons. Because neutrons can form a much more intense ion path as they lose energy within body tissues, neutron radiation is hazardous to body tissues. Other radiations, such as rays, protons, and alpha particles, can be produced as a result of interactions with biological matter. Workers in nuclear power plants and aircraft crews are particularly vulnerable to occupational neutron exposure.

As a result, there is a high demand for effective, long-lasting radioprotective equipment in applications dealing with potential health hazards from various types of radiation. In this section, we will look at radiation and the shielding materials made with boron and boron compounds against it.

The use of special materials such as boron-doped nanoparticles, boron-based polymers, and additives in a boric-oxide matrix to protect people and equipment from

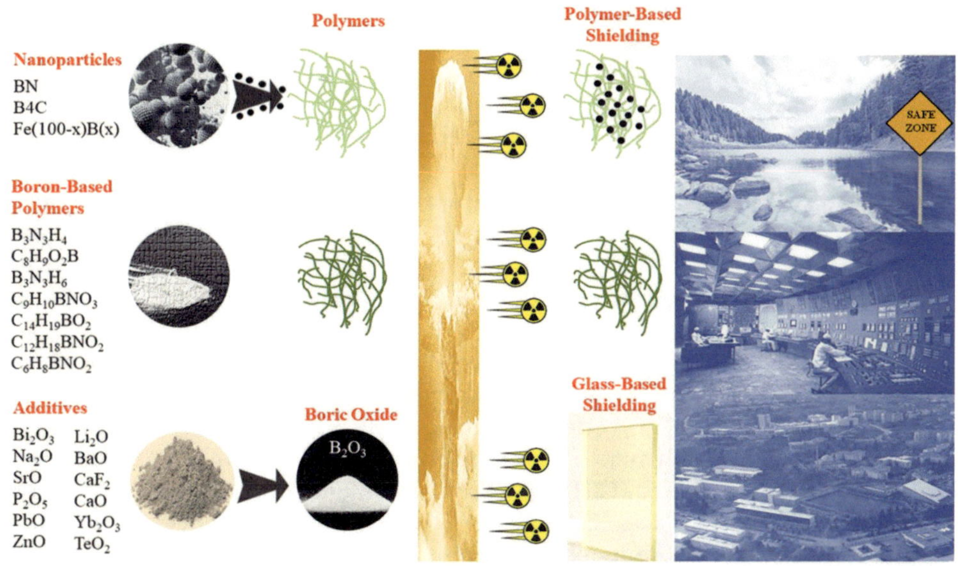

Figure 1.
The boron-doped nanoparticles and boron-based polymers can be effectively used as potential radiation shielding materials in daily life and work-life environments. The use of additives in the boric-oxide matrix is another promising approach for developing glass-based composites for radiation shielding materials.

the harmful effects of ionizing radiation is known as radiation shielding (see **Figure 1**). Ionizing radiation is made up of high-energy particles or electromagnetic waves that can cause harm to biological tissues and other materials. By absorbing, scattering, or blocking radiation, shielding materials can reduce the amount of radiation that reaches a given area. The radiation shield's effectiveness is determined by several factors, including the radiation's energy and type, the thickness and composition of the shielding material, and the distance between the radiation source and the shielding material. The following equation can be used to calculate the amount of radiation passing through a material [4]:

$$I = I_0 e^{-\mu x} \tag{1}$$

where I is the intensity of the radiation after passing through a material, I_0 is the initial intensity of the radiation, μ is the material's linear attenuation coefficient, and x is the material's thickness. The linear attenuation coefficient, which is affected by the energy and type of radiation as well as the material's composition, represents a material's ability to attenuate radiation. The greater the linear attenuation coefficient, the better the material attenuates radiation. The amount of radiation passing through a material is also affected by its thickness; the thicker the material, the more the radiation is attenuated. The equation can be used to calculate the thickness of a shielding material required to reduce radiation to a safe level. The material thickness required to reduce the radiation intensity to the desired level can be calculated using the material's linear attenuation coefficient and the radiation's initial intensity. Radiation shields are materials that are used to protect people and equipment from ionizing radiation. The radiation shield's effectiveness is determined by several factors, including the radiation's energy and type, the thickness and composition of the shielding material, and the distance between the radiation source and the shielding material. Eq. (1) can be used to calculate the intensity of radiation passing through a material and the thickness of shielding material required to reduce the intensity of radiation to a desired level.

2. Gamma, X-ray, and neutron shielding properties of boron and boron compounds

2.1 Gamma, X-ray, and neutron shielding of boron polymers

Nagaraja et al. [5] investigated X-ray and gamma radiation shielding parameters such as mass attenuation coefficient, linear attenuation coefficient, Half Value Layer (HVL), Ten Value Layer (TVL), effective atomic number (Z_{eff}), and electron density of various boron-based polymers [Polymer A-Polyborazilene ($B_3N_3H_4$), Polymer B-4-Vinylphenyl Boronic acid ($C_8H_9O_2B$), Polymer C-Borazine ($B_3N_3H_6$), Polymer D-3-Acrylamidophenylboronic acid ($C_9H_{10}BNO_3$) Polymer E-Phenylethenylboronic acid ($C_{14}H_{19}BO_2$), Polymer F-4-Aminophenylboronic acid ($C_{12}H_{18}BNO_2$) and Polymer G-3-Aminophenylboronic acid ($C_6H_8BNO_2$)]. In addition, the neutron shielding properties of boron polymers were examined. These parameters included the coherent neutron scattering length, the incoherent neutron scattering length, the coherent neutron scattering cross section, the total neutron scattering cross section, and the neutron absorption cross section. They analyzed the different boron polymers' shielding properties and compared them to one another. Based on the findings of the in-depth study, it is clear that the boron polymer phenylethenylboronic acid is an efficient radiation absorber, particularly for X-ray, gamma, and neutron radiation. They concluded that the boron polymer phenylethenylboronic acid is an effective material for shielding X-rays, gamma rays, and neutrons from the environment. They used a NaI(Tl) crystal detector with a detection area of 2.54 × 2.54 cm^2 that was put on a photomultiplier tube that was enclosed in a lead chamber. Additionally, they made use of an advanced PC-based MCA. A powdered form of the compound was placed within a circular holder made of perspex with a diameter of 1 cm and a standard thickness of 1 cm. The substance was attached straightforwardly to the opening in the lead shield that served as the location of the source. During the course of their research, they discovered that the half-value layer and the tenth-value layer of the boron polymer derived from phenylethenylboronic acid were significantly thinner than those of the other boron polymers that were studied. Boron polymer with phenylethenylboronic acid added makes it less permeable to gamma and X-rays compared to other boron polymers with the same composition. It was demonstrated that the boron polymer derived from phenylethenylboronic acid has a shorter mean free path compared to the other boron polymers that were investigated. Boron polymer with added phenylethenylboronic acid makes it less permeable to gamma and X-rays than other boron polymers with the same composition. Nagaraja et al. [5] were able to show that the boron polymer made from phenylethenylboronic acid had a greater effective atomic number compared to the other boron polymers that were investigated. In addition to this, they found that the effective electron density of the boron polymer that was generated from phenylethenylboronic acid had the greatest value of all of the values that were investigated. They studied a variety of boron polymers and compared their lengths, cross sections, total neutron scattering cross sections, and neutron absorption cross sections for coherent and incoherent neutron scattering parameters. For the phenylethenylboronic acid boron polymer, both the coherent neutron scattering length and the incoherent neutron scattering length were found to be at their shortest possible lengths. Phenomenally small coherent and total neutron scattering cross sections characterize the phenylethenylboronic acid boron polymer. A boron polymer that contains phenylethenylboronic acid has been shown to have a significant neutron absorption cross section. They assessed several different metrics of coherent and incoherent neutron scattering for a variety of boron polymers. These metrics included lengths, cross sections, the total neutron scattering cross section, and the neutron absorption cross section. The researchers concluded that the coherent neutron scattering length and the incoherent neutron scattering length for the phenylethenylboronic acid boron polymer were both at their shortest conceivable lengths. This was determined by finding that both lengths were at their smallest possible lengths. Both the coherent and total neutron scattering cross sections are exceedingly low in the case of the boron polymer made from phenylethenylboronic acid. The neutron absorption cross section is relatively high in the case of the phenylethenylboronic acid boron polymer.

2.2 The effectiveness of gamma irradiation of polystyrene-b-polyethyleneglycol-boron nitride (PS-b-PEG-BN) nanocomposites

Cinan et al. [1] wanted to investigate the effectiveness of gamma irradiation and the shielding characteristics of PbO-doped crosslinked PS-b-PEG block copolymers and polystyrene-b-polyethyleneglycol-boron nitride (PS-b-PEG-BN) nanocomposites materials in their work. In order to investigate the gamma-ray shielding properties, crosslinked PS-b-PEG block copolymers and PS-b-PEG-BN nanocomposites were combined with varying percentages of PbO. The production of the copolymer was carried out using various techniques, including emulsion polymerization [6, 7]. For the purpose of their research, the researchers utilized the crosslinked PS-b-PEG block copolymers as a polymeric matrix. They also utilized BN and PbO as the radiation absorption functional material in order to lessen the impact of high-energy gamma rays. The gamma-ray attenuation coefficients were compiled, and a study was conducted using the Linear Attenuation Coefficients (LACs, μ_L) and Mass Attenuation Coefficients (MACs, μ_m) of the crosslinked PS-b-PEG block copolymers-BN-PbO nanocomposites for a variety of items in the linked photon energy area. They found an admissible consistency between the experimental and theoretical μ_L and μ_m of the samples, and the measured and calculated values reflect variations with the modification of the polymer type used to improve the gamma radiation shielding materials. In addition, they found that the samples had an admissible consistency between the experimental and theoretical μ_ms and μ_Ls. In order to achieve the same goal, HVL, TVL, the Mean Free Path (MFP), and the Radiation Protection Efficiency (RPE) values of crosslinked PS-b-PEG block copolymers-BN-PbO nanocomposites were examined in the important critical photon energy area for the gamma-ray attenuation properties. Their findings are a very crucial indicator of the degree to which the material in question is effective at shielding radiation. At the Physics Department of Karadeniz Technical University, the gamma irradiation attenuation factors of the examined composites were obtained for a wide variety of energy spectrums emitted from a ^{152}Eu source using an high-purity germanium (HPGe) detector framework. These energy spectra were measured using gamma rays released from a ^{152}Eu source. To acquire experimental outputs throughout the surveying method, they profited from the computer program Gamma Vision, which maintains powerful multi-channel analyzer capabilities. This allowed them to acquire the outputs of the experiments.

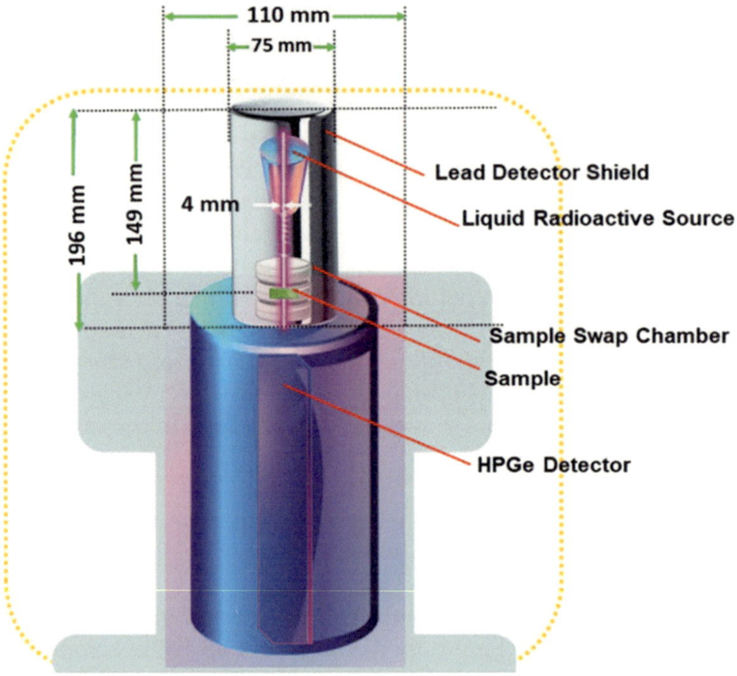

Figure 2.
Diagram of the HPGe detector for gamma irradiation attenuation experiments.

Figure 2 shows the representation of the HPGe detector for gamma irradiation attenuation experiments.

Mixing PbO in the crosslinked PS-b-PEG block copolymers and the PS-b-PEG-BN nanocomposites matrix increases the odds of contact between the incoming gamma irradiation and the shielding material atoms. It can be determined that their samples can treat as shieldings as opposed to the low dosage fractions from gamma irradiation origins. The LACs values fall together with a surge in gamma radiation energies. The occasion for that circumstance is the interaction of gamma radiation with materials *via* a photoelectric effect, Compton scattering, and pair creation. The photoelectric effect is especially significant in the low gamma energy regions; hence, μ_L values are higher in the gamma radiation energy zones in which they are found. The Compton effect is predominant in locations with a medium level of irradiation energy, whereas pair creation is predominant in places with an elevated level of gamma irradiation energy; hence, μ_L worthies begin to decrease with the increase of gamma energies. In addition, it was observed that the radiation protection capacities of the samples improved when PbO or BN percentages of the produced materials were modified. The present study reports on the mass attenuation coefficients (MACs) measured at discrete gamma-ray energy intervals ranging from 121.782 keV to 1408.006 keV. The findings reveal a negative correlation between gamma-ray energy and attenuation aptitude, indicating that as the gamma-ray energy increases, the attenuation of all samples gradually decreases. This is the case. The mixing of PbO in the crosslinked PS-b-PEG block copolymers and the PS-b-PEG-BN nanocomposites matrix results in an increase in the possibility of interaction between the incoming gamma radiation and the gamma-ray shielding atoms. This is because it is more difficult to shield photons with high gamma energies than it is with low gamma energies of photons. It is possible to conclude that the samples can act as shields even when exposed to gamma irradiation sources that produce the modest dose rates. When there is an increase in the amount of gamma radiation present, the values of the μ_ms go down. In addition, it was found that the PbO and BN percentages of the manufactured materials had an effect on the radiation protection capacities of the samples and that these percentages had an improvement in the radiation protection capacities of the samples. When they examine all of the samples side by side, they can conclude that the larger μ_ms and optimal absorption of gamma rays are due to the increased percent rates of the PbO element in the various compounds. It may be underlined that the samples generated in the research disclose significant and dependable results to enlighten radiation shielding studies if the μ_m results of all of the samples are examined with a broad viewpoint. This is because these samples were developed in the course of the research. The HVL, TVL, and MFP values that were obtained for these samples are the most important indications of the radiation shielding ability of the newly created materials for radiation shielding at intervals of gamma-ray energy ranging from 121.782 keV to 1408.006 keV. The value of these coefficients determines how significant their radiation protection efficiency is. The lower the value, the more significant their radiation protection efficiency is. The HVL, TVL, and MFP values all scale up or down about a single property. The TVL value, which is the shielding thickness value required to stop 90% of the emitted photons, that is, to absorb them, increases as the photon energy increases. This is because the shielding thickness value is proportional to the value required to stop 90% of the photons. Their findings are an extremely useful indicator of the radiation shielding capacity of the material in question; more specifically, one may conclude that the lower the TVL value of any given sample is, the greater the radiation shielding efficiency will be as a result of the reduced thickness requirements. The HVL values of the samples provide convincing guidance regarding the shielding capacity of the materials in decreasing the photon quantities to half of what they are currently at for the sample thickness. One of the main characteristics that clearly explain the gamma radiation degrading abilities of the shielding substances that are used is the MFP value. The better a substance's ability to shelter other particles from radiation, the lower the MFP value should be. The results that they obtained show that the MFP rates of the other samples increase as the photon energy increases. The success of the materials that were created to measure the attenuation of the gamma photons in the various energy intervals can be monitored by computing the RPE. Therefore, the RPE values for the PbO doped the crosslinked PS-b-PEG block copolymers and the PbO-doped PS-b-PEG-BN nanocomposite materials have been monitoring the densities of the photons as a function of different gamma-ray energy intervals. In their research, it can be seen that the RPE values tend to decrease with increasing energy for all the evaluated composites. According to these findings, the PbO-doped polymer-based composites that they developed have a good performance in shielding gamma radiation. That is to say, the recent adjustments to the doping ratios are proving to be quite successful in lowering the intensity of gamma photons. In addition, it was found that the PbO and BN percentages of the manufactured materials had an effect on the radiation protection capacities of the samples and that these percentages had an improvement in the radiation protection capacities of the samples. When one looks at the HVL and TVL values of those samples, one can see that the sample with the PS-PEG (1000)-BN-S0 has the best HVL value for 121.782 keV with 1.336 cm, and the sample with the TVL value of 4.439 cm has the best value for 121.782 keV. It can be seen that the PS-PEG (1500)-BN-S0 sample has the best HVL value, which is 7.801 cm, and the best TVL value, which is 25.913 cm. Both of these values come from the energy level 1408.006 keV. When they compare these values with previously indicated values, which are solely polymers in their structure, the contribution of adding BN to the composite is visible. In addition, BN is thought to be a good neutron absorber. Based on this concept, it is simple to conclude that the contribution of BN is significant, given that one expects that the mineral that was added will also be useful for neutron radiation. This conclusion can be reached since one thinks that the mineral will be useful for neutron radiation. As a consequence of this, the materials from their investigation show that PbO doping occurred in the crosslinked PS-b-PEG block copolymers and PbO doping occurred in the PS-b-PEG-BN. Nanocomposite materials are excellent choices for achieving radiation protection objectives for gamma rays. These materials are particularly advantageous as a shielding substance for transporting radiation sources and as an insulating substance for radioactive waste administration facilities or the building industry. When it comes to boosting the hardness, durability, and radiation absorption capacities of the shielding materials, the connecting of several sorts of contributions (such as cement, polymer, and metal oxide, among others) is of the utmost importance. As a result of their low cost and low weight, polymer structures are a significant class of substances that are utilized in radiation shielding research. In addition, polymer structures will be the starting point for many different types of research utilizing composites acquired by suffixing micro or nano-oxide, etc., to investigate radiation attenuation both theoretically and experimentally.

2.3 The gamma-ray shielding properties of the polymer-nanostructured selenium dioxide (SeO_2) and boron nitride (BN) nanoparticles

The gamma-ray shielding properties of crosslinked PS-b-PEG block copolymers combined with nanostructured SeO_2 and BN nanoparticles were investigated by Cinan et al. [2] in their work. The PS-b-PEG copolymer as well as nanostructured SeO_2 and BN particles, all had a substantial impact on the enhancement of the resistance of the nanocomposites, and the samples with high additive rates demonstrated superior resistance than the other nanocomposites. As a result of the accomplishments, it is possible to conclude that the polymer-based nanocomposites can be utilized as a viable option in the gamma-irradiation-shielding sector of the industry. Their nanocomposites' irradiation properties were studied using rays from a ^{152}Eu source in an HPGe detector setup, and the results were evaluated using Gamma Vision software. In addition, the theoretical calculus was used to determine all of the radiation shielding factors, and these were compared to the findings of the experiments. Because different rays of different energy and wavelengths have varied interactions with the atoms in the material, the ^{152}Eu radioactive source was utilized to offer the most thorough data. The comparability between the experimental findings and the theoretical predictions was found to be satisfactory in all of the nanocomposites.

The PS-b-PEG copolymer as well as nanostructured SeO_2 and BN particles had a key role in the enhancement of the resistance of the nanocomposites, and the samples with high additive rates displayed superior resistance than the other nanocomposites did. **Figure 3** presents SEM images of the crosslinked PS-b-PEG block copolymer with BN nanocomposites. BN nanoparticles mixed homogeneously with PS-b-PEG block copolymer. There are roughnesses, particles, pores, and elevations on the surface. Based on the results that were obtained, it is possible to conclude that polymer-based nanocomposites are a good option for use in the gamma-irradiation-shielding

Figure 3.
SEM photographs of the crosslinked PS-b-PEG (10,000) block copolymer+BN nanocomposite (50% PS-PEG (10,000)+50% BN+0% PbO): (a) PS-PEG (10,000)-BN-S0 (12,000× magnification); (b) PS-PEG (10,000)-BN-S0 (4480× magnification).

discipline in many different applications (such as flexible and durable gamma-radiation-protective systems for the transportation of radioactive materials, isolation for the operations of radioactive waste, and radiation services in hospitals, nuclear power plants (NPPs), the defense industry, the building industry, and many other applications). The theoretical and experimental mL values of the studied nanocomposites were found to be in satisfactory concordance with one another. In particular, the PS-b-PEG block copolymers blended with the nanostructured SeO_2 and BN particles' nanocomposite matrix resulted in a significant enhancement in the possibilities of reciprocal influence between the arriving gamma rays and the shielding nanocomposite atoms. This was the case since the PS-b-PEG block copolymers were blended with the nanocomposite matrix. They can conclude that the nanocomposites they have researched can also be employed as materials that guard against low and large doses of gamma radiation. They demonstrated that the μ_L rates tended to fall when the gamma energy was increased. It was determined that the experimental and theoretical μ_L rates exhibited good harmony and improved shielding behavior with the change in the polymer type utilized to manufacture gamma-ray-absorbing nanoparticles. This was one of the conclusions that were reached. Moreover, they discovered that the gamma-ray protection properties of the nanocomposites increased when the amounts of nanostructured SeO_2 and BN particles contained in the nanomaterials were adjusted. This was another finding made by the researchers. The μ_m values of the PBSNC5, PBSNC9, PBSNC6, PBSNC8, and PBSNC12 nanocomposites [2] were lower than the mm values of their respective copolymers when the composites including copolymers, SeO_2, and BN nanoparticles were tested at 121.782 keV. These nanocomposites had μ_m values of 0.151, 0.142, 0.242, 0.263, and 0.329, respectively. At this energy, the mm values of the PSNC6 and PSNC11 copolymers [2] dropped from 0.283 to 0.151 and from 0.177 to 0.142, respectively, after the addition of 50 wt% nanostructured BN to the PSNC6 and PSNC11 copolymers. This change was caused by the addition of nanostructured BN. It was determined that the experimental and theoretical μ_m rates exhibited good harmony; more specifically, it was determined that the PS-b-PEG copolymers combined with nanostructured SeO_2 and BN particles in a nanocomposite matrix showed good attenuation and protective outcomes against gamma irradiation. It is possible to conclude that the nanocomposites have the potential to be used as shielding materials against low and high gamma doses in a variety of settings. For all of the nanocomposites' outcomes, it has been thoroughly emphasized that the experimental and theoretical values demonstrate changes when the type of copolymer that was used to build the nanomaterial for the attenuation of and protection against the impacts of gamma rays is changed. These variations can be seen in both the attenuation and protection that the nanomaterial provides. In addition to this, it was observed that the radiation shielding performance of the other nanocomposites increased when the nanostructural proportions of SeO_2 or BN particles included in the nanocomposites were raised. When the μ_L and μ_m values of all of the nanocomposites are examined, it is clear that the nanostructured composites cultured in this work demonstrate significant and reliable outcomes in terms of radiation absorption and protection. These findings can be seen when the nanocomposites are examined.

The HVL, TVL, MFP, and RPE rates that have been determined for the nanocomposites are the factors that have the largest impact on the gamma-shielding effectiveness. The gamma-ray shield's qualities have a greater influence on the environment when the HVL, TVL, and MFP rates are reduced. In addition, the performance of the polymer-based nanocomposites that were manufactured to develop the shield for protection against gamma rays over a wide energy range may be seen by calculating the RPE rates. This will show the performance of the nanocomposites. Additionally, it was discovered that the gamma irradiation protective properties of the nanocomposites rose when the amounts of nanostructured SeO_2 and BN particles present in the nanomaterials were increased. This was another finding that was made. The TVL rates of the PS-b-PEG block copolymers blended with the nanostructured PbO particles changed from 5.799 cm to 30.725 cm. On the other hand, the TVL rates of the PS-b-PEG (1000) block copolymer blended with the nanostructured SeO_2 particles did not change. Furthermore, the HVL rates of the PS-b-PEG copolymers blended with the nanostructured PbO and BN particles changed from 0.967 cm to 7.347 cm (for 15% PS-b-PEG (10,000) copolymer, 15% nanostructured BN, and 70% PbO particles) [8], and the HVL values of the PS-b-PEG (10,000) block copolymer blended with the nanostructured SeO_2 and BN particles changed from 0.843 cm to 7.203 cm (for 15% PS-b-PEG (10,000) copolymer, 15% nanostructured BN, and 70% SeO_2 nanoparticles). To be more specific, the nanostructured SeO_2 additive decreased the thickness while simultaneously improving the radiation absorption efficiency. In addition, the ability of the polymer-based nanocomposites that were designed to detect the attenuation behaviors of the gamma rays in a wide energy range can be detected by computing the RPE rates. When confronted by gamma rays, the copolymers that were combined with nanostructured SeO_2 and BN nanocomposites had excellent shielding efficacy (see **Figure 4**). This conclusion may be drawn from all of the researchers' findings. The shielding properties of these composites were comprehensively obtained with the ^{152}Eu gamma radioisotope source by detecting the outputs of nanostructured SeO_2 and BN particles blended or unblended in PS-b-PEG-based composite tablets with the experimental system. The infrastructure of this system was created with theoretical calculations *via* the radiation parameters. In addition to this, the morphological and temperature degradation features were examined. It has been observed that increasing the amount of nanostructured SeO_2 and BN particles in PS-b-PEG copolymer-structured composite materials results in an increase in the amount of radiation shielding and protection from dangerous gamma rays given by the materials. In this context, the behavior of SeO_2 and BN blended and unblended nanocomposites against a gamma radioisotope source with a wide energy range was

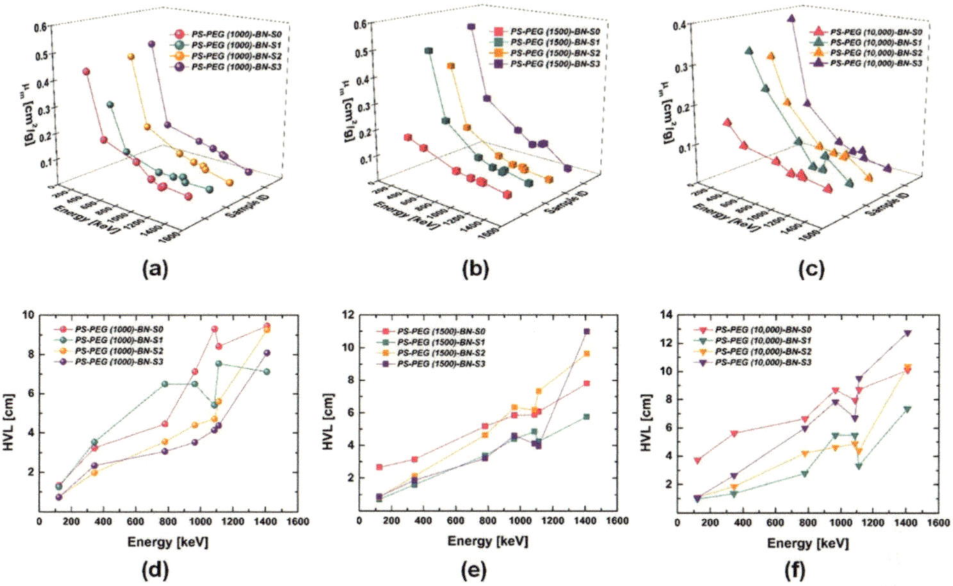

Figure 4.
The μ_m rates and HVL values of the PS-b-PEG copolymers blended with the nanostructured SeO2 and BN particles under a wide range of gamma irradiation energies.

investigated. Additionally, an application-oriented study that can be used in fields such as nuclear technology was carried out, and the results of this study can contribute to the scientific literature. The TEM photos also revealed another significant finding, which was that the addition of BN nanoparticles to the nanocomposite brought about a discernible shift in the distribution as well as the particle structure of the SeO_2 nanoparticles present in the composite.

2.4 Design and fabrication of high-density borated polyethylene nanocomposites as a neutron shield

Mortazavi et al. [9] have shown that neutron shielding using polyethylene composites containing boron can be accomplished in a very efficient manner. Their investigation is centered on the manufacturing of borate polyethylene nanocomposites. The purpose of this research is to develop a radiation shield that can be employed effectively in situations in which the user is subjected to both neutron and gamma radiation. They started by making borate polyethylene shields that had 2 and 5% by weight of boron nanoparticles, and then, they compared the neutron attenuation of those shields to that of pure polyethylene. In order to determine the amount of attenuation that Am-Be neutrons experience when traveling through the shields, they used polycarbonate sheets. The mean (standard deviation) of the number of traces induced by neutrons traveling through shields was 1.048810 3 128.98 for polyethylene with 5% by weight, and 289.5610 3 1.1972, and 1.534010 3 206.52 for polyethylene with 2% by weight boron nanoparticles and pure polyethylene. The tensile strength of borate polyethylene nanocomposites was found to be greater than that of pure polyethylene. Neutron attenuation was compared between a borate polyethylene nanocomposite that had 5% by weight boron in it and pure polyethylene, and the results showed that there was a statistically significant difference between the two.

However, there was not a statistically significant difference between having 5% by weight of boron borate in a polyethylene nanocomposite and having 2% by weight of boron. It was also established in this research that the tensile strengths of boron carbide nanocomposites are significantly greater than those of pure polyethylene. It is important to note that the findings of Mortazavi et al.'s earlier research on photon shielding show that the use of nano-sized materials in radiation shielding can only provide better attenuation results in very particular circumstances, such as a limited photon energy range or a limited concentration of nanomaterials in the matrix. This is something that should be kept in mind.

2.5 The nuclear shielding of iron-boron alloys

Iron-boron alloys play a significant role in powder metallurgy, and Aytac and coworkers [10] looked at their nuclear radiation shielding properties for this study. When bombarded with ^{152}Eu, Fe(100-x)B(x) alloys (where x is 1, 2, 4, 6, 8, 10, 12, 14, 16, 18, and 20) produce photons with energies of 0.0810, 0.1218, 0.2764, 0.3029, 0.3560, 0.3443, 0.3839, and 0.7789 MeV, respectively. The photon intensities were measured with an Ultra-Ge detector. Half-value layers, mean free routes, effective atomic numbers, and effective electron densities were all computed using experimentally observed μ_m. The results demonstrate that the HVL and MFP values are best for the Fe-B alloy with 20% boron. Research has demonstrated that the addition of boron does not improve Fe-B alloys' photon-shielding properties. However, it was shown to be quite effective in terms of gamma attenuation in the chosen energy range when compared to previously reported shield materials. Neutron dose transmission studies were conducted as well, with removal cross-sectional values ($\sum R$) determined. In contrast to their gamma-shielding properties, Fe-B alloys' neutron reduction capability increased with increasing boron content. The findings from each of the gamma shielding parameters demonstrated that an increasing quantity of boron had a detrimental effect on the alloys' capacity to cut down on gamma radiation. In addition to that, measurements of the equivalent neutron dosage were carried out, and the effective removal cross sections of the alloys were gathered. Based on these factors, it was determined that the alloy's neutron-holding capacity rose along with the percentage of boron present in the alloy. It is anticipated that as a consequence of this, it will be possible to conclude that Fe-B alloys are more effective at absorbing neutron radiation than gamma radiation.

2.6 Boron nitride nanosheet-reinforced WNiCo-FeCr HEAs

WNiCo-FeCr high entropy alloys (HEAs) were the subject of research conducted by Kavaz et al. [11], who investigated the synthesis and complete characterization of these alloys. These alloys were reinforced with newly created boron nitride nanolayers (BNNSs). In this work, a comprehensive investigation was conducted into the effect of B_4C on the structural, physical, mechanical, and nuclear protective features of synthesized HEAs. The investigation focused on the effect of B_4C's monotonous behavior modifications. They were able to ascertain the characteristics of protection against nuclear radiation through the use of experimental gamma-ray and neutron assemblies. In addition, the properties of nuclear radiation shielding for gamma rays and fast neutrons were carefully compared with the properties of many other kinds of shielding materials, both those already in use and those of the next generation. They concluded that an elevated level of B_4C directly adds to the shielding qualities of nuclear radiation after reviewing the relevant evidence. They state that the B_4C that is formed within the structure of BNNS will contribute to the overall properties of HEAs. This is highly essential for nuclear applications, as HEAs are now being examined as a component of potential future nuclear reactors. In addition, they concluded that B_4C is a versatile material that can be utilized in settings where the mechanical and nuclear shielding qualities need to be improved for a variety of radiation intensities. This led them to the conclusion that B_4C is a versatile material.

2.7 B_4C particle-reinforced Inconel 718 composites

Gokmen [12], in his work, wanted to offer a computational tool that would perform calculations of critical physical variables for the gamma-ray attenuation in the B_4C (0.25 wt%) particle-reinforced Inconel 718 superalloy. Specifically, the purpose of this study was to determine the gamma-ray attenuation in the material. It was the first time that the shielding qualities of the B_4C (0.25 wt%) particle-reinforced Inconel 718 superalloy were investigated for use in nuclear technology as well as other technologies such as nanotechnology and space technology. The relationship between the weight percentage of the B_4C component in these superalloy materials and the attenuation of gamma rays was investigated. The LAC, MAC, Exposure Buildup Factors (EBF), TVL, HVL, MFP, Z_{eff}, and fast neutron removal cross sections (FNRC) values of the B_4C (0e25 wt%) particle-reinforced Inconel 718 superalloy composites (which contains special alloy elements such as Ni, Cr, Nb, Mo) were theoretically calculated for the first time in order to evaluate the effectiveness of gamma and neutron radiation shielding using the PSD software. The attenuation performance of the materials, which may be utilized as shielding materials, improved as a result of a decrease in the weight fraction of the B_4C compound in the Inconel 718 superalloy composites, as evidenced by the computed values of the MAC. The MFP, HVL, and TVL values were found to increase as the gamma-ray energy and the weight percent of the B_4C component found in the Inconel 718 superalloy composites was increased. This was observed to be the case. This occurred because of the density as well as differences in the chemical makeup of certain superalloy compounds. Due to the greater values of MFP, HVL, and TVL, it can be deduced that thicker materials are necessary to attenuate radiation to a level that is considered safe. Because of this, it is preferable to choose materials with lower HVL values rather than those with higher HVL values in order to cut down on both the cost and the size. As a consequence of this, the Inconel 718 superalloy materials with B_4C addition are incapable of successfully absorbing a wide variety of gamma rays. This is because these materials have a low gamma absorption cross section. Because of this, the superalloy Inconel 718 material known as S6 which contains B_4C in a weight percentage of 25% was discovered to be the most effective neutron shielding material. Because the energy of the incident gamma rays influences both the selection of the shielding material and the determination of the required thickness of the shielding material, this study reveals that each quantity is dependent on the energy.

2.8 Borate glasses

The expanding applications of gamma radiation in fields such as health, industry, and agriculture call for the research and development of radiation shielding materials that are see-through. According to Kirdsiri et al. [8], glass materials are perfect for this purpose since they can be recycled completely, they do not have an opaque appearance, and they can be modified and transformed by adding new components.

Nuclear radiation attenuation and mechanical characteristics were inspected by Lakshminarayana et al. [13] for 10 lithium bismuth borate glasses with varying levels of Bi_2O_3 concentration (ranging from 10 to 55 mol% and denoted as glasses A to J). This was done in order to gain an understanding of how these properties would change with an increase in Bi_2O_3 content. At a selection of twenty-five energies ranging from 15 KeV to 15 MeV, photon shielding abilities in respect to μ, μ/ρ, Z_{eff}, N_{eff}, HVL, TVL, MFP, and RPE have been investigated. At each energy level, it was found that the μ/ρ values that were calculated using theoretical (Py-MLBUF [14] and WinXCOM [15]) and computational (MCNPX [16], FLUKA [17, 18], and PHITS codes [19]) methods were in qualitative agreement with one another. For example, each Py-MLBUF, WinXCOM, MCNPX, FLUKA, and PHITS algorithms got a different /g value for sample J at 0.6 MeV energy. These values were 0.119, 0.1186, 0.1126, 0.1183, and 0.1174 cm^2/g, respectively.

In this research work, Almuqrin et al. [20] aim to evaluate the radiation shielding capabilities of a Yb^{3+}-doped calcium borotellurite glass system as part of their investigation. The system's fundamental components are CaF_2–CaO–B_2O_3–TeO_2–Yb_2O_3, but for convenience's sake, one is referring to it as TeBYbn. It was determined what would happen if the amount of TeO_2 in the glasses was increased from 10 to 54 mol% by experimenting with five distinct combinations of compositions and densities. Investigation into the μ_m (μ/ρ) of the samples was carried out with the help of the Phy-X/PSD program [21]. The mass attenuation coefficients were calculated theoretically by using an online program that was designed to calculate shielding characteristics. Other metrics, such as the μ_L, transmission factor (TF), RPE, Z_{eff}, and MFP, were then computed and studied after that. TeBYb5, the glass with the highest TeO_2 content, was demonstrated to have the highest μ/ρ; however, at higher energies, the variations between the values became essentially insignificant. It was discovered that density increases with density, such as going from 0.386 to 0.687 cm^{-1} for TeBYb1 and TeBYb5 at 0.284 MeV, respectively. It was determined that there is an inverse association between the thickness of the sample and the TF because it was discovered that samples with a thickness of 1.5 cm had the lowest TF. Because both the HVL and the TVL of the samples declined as the density of the samples dose, one can conclude that TeBYb1 is the least effective of all of the glasses that were studied. These five samples were able to demonstrate their effectiveness as radiation shields by having an MFP that was lower than that of certain other types of shielding glasses. TeBYb5 appeared to have the greatest capacity to attenuate photons based on the parameters that were determined.

Lakshminarayana et al. [22] investigated the gamma-ray and neutron attenuation properties of both the B_2O_3-Bi_2O_3-CaO and the B_2O_3-Bi_2O_3-SrO glass systems in their research. This was done for both of the glass systems. Within the energy range of 0.015 to 15 MeV, linear attenuation coefficients (μ) and mass attenuation coefficients (μ/ρ) were estimated by using the Phy-X/PSD program. The obtained numbers fit up quite well with the respective simulation results computed by the MCNPX, Geant4 [23–25], and Penelope [26] programs. The inclusion of Bi_2O_3, rather than B_2O_3/CaO or B_2O_3/SrO, results in increased gamma-ray shielding competency. This is indicated by an increase in the Z_{eff} as well as a decrease in the HVL, TVL, and MFP. Within the range of 0.015 to 15 MeV, a geometric progression (G-P) fitting approach was utilized to determine EBFs and energy absorption buildup factors (EABFs) at 1 to 40 mfp penetration depths (PDs). The RPE values that have been computed show that they have a high capacity for shielding photons with lower energies, comparatively greater density (7.59 g/cm^3), larger μ/ρ, Z_{eff}, equivalent atomic number (Z_{eq}), and RPE, along with the lowest HVL, TVL, MFP, EBFs, and EABFs derived for $30B_2O_3$-$60Bi_2O_3$-10SrO (mol%) glass, implying that it could be an excellent gamma-ray attenuator. Additionally, $30B_2O_3$-$60Bi_2O_3$-10SrO (mol%) glass holds a commensurably bigger macroscopic removal cross section for fast neutrons. $\sum_R = 0.1199\, cm^{-1}$ obtained by applying Phy-X/PSD for fast neutron shielding, owing to the presence of a larger wt% of 'Bi' (80.6813 wt%) and moderate "B" (2.0869 wt%) elements in it. Because it has a high weight percentage of the "B" element, the sample with the composition $70B_2O_3$-$5Bi_2O_3$-25CaO (mol%) (B: 17.5887 wt%, Bi: 24.2855 wt%, Ca: 11.6436 wt%, and O: 46.4821 wt%) has a high potential for the capture or absorption of thermal or slow neutrons and intermediate energy neutrons.

During the course of their research, Madbouly and colleagues [27] looked at bismuth-borophosphate glasses. One type of glass known as borophosphate glass contains phosphorus and boron oxide as components. Borophosphate glass is an important category of glass that can be distinguished by several useful qualities that it contains. Borophosphate glass has been demonstrated to have excellent ionic conductivity, and its preparation is very straightforward. Despite its low melting point and high glass-forming capacity, the application of borophosphate glass has been limited because of its hygroscopic character. This is even though it has a relatively low melting point. It is recommended that a heavy metal oxide (HMO) be added to the produced glass in order to increase its potential for radiation shielding. The metal oxide with the highest density is known as HMO. In addition to this, the high atomic number of Bi contributes to the enhanced gamma shielding capabilities of the glass [28]. In applications that make use of radiation, protective glasses are increasingly being utilized to absorb incoming photons that could potentially harm employees and patients in the area surrounding the radioactive source. Radiation is currently being utilized in hundreds of applications spanning a variety of industries, such as the medical area and the production of energy. Even though radiation has certain positive effects, extra caution is required while working with radioactive sources since photons with a high energy level pose a significant threat to the human body. In this work, they investigate the effect that Bi_2O_3 has on the structure of borophosphate glasses as well as the optical and radiation shielding properties of these glasses. They measured the photon transmissions, μ_Ls, HVL, TVL, and MFP values of bismuth-borophosphate glasses experimentally. The gamma-ray energies they employed were 662, 1173, 1275, and 1333 keV. After that, the results of the measurements were checked against the FLUKA code. The conclusions from the FLUKA code were in good agreement with the results of the experiments. In addition, the data demonstrate that increasing the amount of Bi_2O_3 in the glass network improves the quality of shielding. According to the data they currently have, the absorbance increases in tandem with a rising Bi_2O_3 level. Bismuth-borophosphate glasses have excellent gamma-ray shielding capabilities, making them a good choice for shielding applications.

For shielding applications against the energies emitted by ^{22}Na and ^{131}I isotopes, Al-Buriahi [29] used FLUKA to examine the radiation of gamma and fast-neutron shielding performance of borate glasses containing zinc, bismuth, and lithium (as modifier). Glasses ranging from 0 to 20 mol% xBi_2O_3-(25-x) Li_2O-$60B_2O_3$-15ZnO were analyzed. For the ^{22}Na isotope, the simulations are in run for energies of 0.511 and 1.275 MeV, and for the ^{131}I isotope, the energies of 0.365, 0.637, 0.284, and 0.723 MeV. In addition, the report explored the borate glasses' capacity to deflect fast neutrons, thermal neutrons, and charged particles. The results show that the ^{22}Na isotope emits photons with energies as low as 0.284 MeV and that the borate glasses of

BLBZ1, BLBZ2, BLBZ3, BLBZ4, and BLBZ5 have values of 0.282, 0.349, 0.887, 1.103, and 1.397 cm^{-1} and values of 0.108, 0.198, 0.251, 0.286, and 0.310 cm^2/g. At 0.2 MeV, the borate glasses BLBZ1, BLBZ2, BLBZ3, BLBZ4, and BLBZ5 had maximum total SP of electron interaction values of 2.292, 2.076, 1.949, 1.865, and 1.807 MeV cm^2/g, respectively. Furthermore, for the current borate glass samples, the effect of Bi_2O_3 content on the dosage rate was quite minimal. The removal cross sections for fast neutrons in the borate glasses of BLBZ1, BLBZ2, BLBZ3, BLBZ4, and BLBZ5 were 0.112, 0.111, 0.103, 0.100, and 0.106 cm^{-1}, respectively. It is concluded that BLBZ5 glass has the potential to be developed as a viable option for gamma applications. **Table 1** shows the glass code, chemical formula, and weight fraction (wt %) for each component in the current glass specimens.

Borotellurite glasses with the molar compositions of $60TeO_2$–$20B_2O_3$–(20-x)Bi_2O_3–xPbO, where x = 0, 1, 2, 3, 4, 5, 6, 7, 8, 9, and 10 mol%, were synthesized with the help of a typical melt quenching procedure and investigated in terms of their physical, optical, structural, and gamma shielding capabilities. Marzuki and colleagues [30] researched these spectacles. After being mixed, the powder was subsequently melted in an electric furnace at a temperature of around 1000°C for approximately 60 minutes. After being quenched by pouring it into a preheated parallel plate brass mold, the molten metal was subsequently annealed at 350°C for roughly 180 minutes before being cooled to room temperature at a rate of 1°C per minute. When the concentration of PbO is increased from 0 to 10 mol%, the density of the current samples decreases from 6.08 to 5.93 g/cm^3, and the molar volume decreases from 33.37 to 30.27 cm^3/mol. This is because there is a decrease in the concentration of the element Bi_2O_3. Because PbO is used throughout the network rather than Bi_2O_3, the optical packing density has increased from 71.91 to 72.67% as a direct result of this change. When the concentration of PbO increases, the refractive index, molar refractivity, and ionic polarizability all decrease. Specifically, the refractive index drops from 1.9137 to 1.8306, the molar refractivity drops from 18.531 to 16.884 cm^3, and the ionic polarizability drops from 7.3534 to 6.7000 3. An increasing fraction of the glass network will be formed of [TeO_3] (tp) and [TeO_3+1] polyhedra as the amount of lead oxide present in the glass increases. The Phy-X PSD software [21] was used in the theoretical research of the gamma radiation shielding qualities, and the photon energy was varied from 0.015 to 15 MeV throughout the study. When considering all glasses, the values of the μ_L that are observed to be the highest are found at 0.015 MeV, while the values that are observed to be the values that are observed to be lowest are found around 5 MeV. At this stage of the gamma photon-electron interaction, the formation of pairs starts to

Glass code	Chemical formula	wt%				Bi
		Li	B	O	Zn	
BLBZ1	$60B_2O_3$-15ZnO-$25Li_2O$	0.05648	0.21112	0.57280	0.15960	
BLBZ2	$60B_2O_3$-15ZnO-$20Li_2O$-$5Bi_2O_3$	0.03335	0.15583	0.44200	0.11780	0.25102
BLBZ3	$60B_2O_3$-15ZnO-$15Li_2O$-$10Bi_2O_3$	0.01982	0.12349	0.36550	0.09335	0.39784
BLBZ4	$60B_2O_3$- 15ZnO-$10Li_2O$-15 Bi_2O_3	0.01094	0.10226	0.31529	0.07731	0.49420
BLBZ5	$60B_2O_3$-15ZnO-$5Li_2O$-$20Bi_2O_3$	0.00467	0.08726	0.27981	0.06597	0.56229

Table 1.
Glass code, chemical formula, and weight fraction (wt%) for each component in the current glass specimens.

take precedence over everything else. When there is an increase in the concentration of PbO, there is a corresponding decrease in the concentration of Bi_2O_3; this causes the LAC values to decrease across all photon energies. The values of the μ_L change from a high of 413.2183 cm^{-1} to a low of 12.4962 cm^{-1} when the potential energy is 0.015 MeV. This dependence on composition is at its most pronounced for photon energy in the range of 0.015–0.04 MeV. When the energy is greater than 0.04 MeV, this dependence is greatly diminished. The outcomes of this experiment at μ_L suggest that using Bi_2O_3 as a gamma shielding material rather than PbO is likely to give better results. The experiment was conducted to investigate this hypothesis.

The study conducted by Mahmoud et al. [31] sought to examine the impact of Pr^{3+} ions on the structural, optical, and gamma-ray shielding characteristics of borosilicate glasses. This study produced a set of borosilicate glass specimens comprising five distinct samples. The samples were composed of a mixture of (55-x) B_2O_3, $15SiO_2$, 20CaO, $10Li_2O$, and xPr_6O_{11}. The objective of the study was to investigate the effects of Pr_6O_{11} on the structural, optical, and X-ray shielding characteristics of borosilicate glass. The findings obtained from the UV-Vis IR spectrometer indicate a reduction in the direct energy gap from 3.508 to 3.304 eV, which is accompanied by an elevation in the Urbach energy from 0.335 to 0.436 eV. These changes are observed as the concentration of Pr_6O_{11} is increased from 0 to 1 mol%. Furthermore, the Monte Carlo simulation outcome about the shielding characteristics of X-rays illustrates an increase in the micrometer values as a result of the minor amounts of Pr_6O_{11} reinforcement. At an energy level of 103 keV, the micrometer values experienced a 161% increase, while the 0.5 values decreased by 62% upon increasing the Pr_6O_{11} concentration from 0 to 1 mol%. At higher intermediate and high-energy gamma (E) values, the gamma-ray shielding properties of the produced glasses exhibit a negligible improvement. This study posited that the samples under investigation are suitable for shielding photons possessing energies that are less than 511 keV.

3. Conclusions

Nanocomposite materials are excellent choices for achieving radiation protection objectives for gamma rays. These materials are particularly advantageous as a shielding substance for the transportation of radiation sources and as an insulating substance for radioactive waste administration facilities or the building industry. As a result of their low cost and low weight, polymer structures are a significant class of substances that are utilized in radiation shielding research. A boron polymer that contains phenylethenylboronic acid has been shown to have a significant neutron absorption cross section. They assessed several different metrics of coherent and incoherent neutron scattering for a variety of boron polymers. In addition, polymer structures will be the starting point for many different types of research utilizing composites acquired by suffixing micro- or nano-oxide, etc., to investigate radiation attenuation both theoretically and experimentally. It has been observed that increasing the amount of nanostructured SeO_2 and BN particles in PS-b-PEG copolymer-structured composite materials results in an increase in the amount of radiation shielding and protection from dangerous gamma rays given by the materials. In this context, the behavior of SeO_2 and BN blended and unblended nanocomposites against a gamma radioisotope source with a wide energy range was investigated. It was also established that the tensile strengths of boron carbide nanocomposites are significantly greater than those of pure polyethylene. The findings from each of the gamma shielding parameters demonstrated that an increasing quantity of boron had a detrimental effect on the alloys' capacity to cut down on gamma radiation. In addition to that, measurements of the equivalent neutron dosage were carried out, and the effective removal cross sections of the alloys were gathered. Based on these factors, it was determined that the alloy's neutron-holding capacity rose along with the percentage of boron present in the alloy. It is anticipated that as a consequence of this, it will be possible to conclude that Fe-B alloys are more effective at absorbing neutron radiation than gamma radiation. It is concluded that B_4C is a versatile material that can be utilized in settings where the mechanical and nuclear shielding qualities need to be improved for a variety of radiation intensities. This led them to the conclusion that B_4C is a versatile material. The Inconel 718 superalloy materials with B_4C addition are incapable of successfully absorbing a wide variety of gamma rays. This is because these materials have a low gamma absorption cross section. Because of this, the superalloy Inconel 718 material known as S6 which contains B_4C in a weight percentage of 25% was discovered to be the most effective neutron shielding material. Because the energy of the incident gamma rays influences both the selection of the shielding material and the determination of the required thickness of the shielding material, this study reveals that each quantity is dependent on the energy. The boron-doped nanoparticles and boron-based polymers can be effectively used as potential radiation shielding materials in daily life and work-life environments. The use of additives in the boric-oxide matrix is also another promising approach for the development of glass-based composites for radiation shielding materials.

Acknowledgements

The authors thank to Karadeniz Technical University for their support. In addition, the authors also would like to express gratitude to Bilkent University UNAM for their kind hospitality.

Author details

Ahmet Hakan Yilmaz[1*], Bülend Ortaç[2] and Sevil Savaskan Yilmaz[2,3]

1 Department of Physics, Faculty of Sciences, Karadeniz Technical University, Trabzon, Turkey

2 UNAM-National Nanotechnology Research Center and Institute of Materials Science and Nanotechnology, Bilkent University, Ankara, Turkey

3 Department of Chemistry, Faculty of Sciences, Karadeniz Technical University, Trabzon, Turkey

*Address all correspondence to: hakany@ktu.edu.tr

References

[1] Cinan ZM, Erol B, Baskan T, Mutlu S, Savaskan Yilmaz S, Yilmaz AH. Gamma irradiation and the radiation shielding characteristics: For the Lead oxide doped the crosslinked polystyrene-b-Polyethyleneglycol block copolymers and the polystyrene-b-Polyethyleneglycol-boron nitride nanocomposites. Polymers. 2021;**13**:3246-3276. DOI: 10.3390/polym13193246

[2] Cinan ZM, Erol B, Baskan T, Mutlu S, Savaskan Yilmaz S, Yilmaz AH. Radiation shielding tests of crosslinked polystyrene-b-Polyethyleneglycol block copolymers blended with nanostructured selenium dioxide and boron nitride particles. Nanomaterials. 2022;**12**:297-328. DOI: 10.3390/nano12030297

[3] More CV, Alsayed Z, Badawi MS, Thabet AA, Pawar PP. Polymeric composite materials for radiation shielding: A review. Environmental Chemistry Letters. 2021;**19**:2057-2090. DOI: 10.1007/s10311-021-01189-9

[4] Demir N, Kuluozturk ZN, Dal M, Aygün B. Investigation of gamma-ray shielding parameters of marbles. In: Mann KS, Singh VP, editors. Computational Methods in Nuclear Radiation Shielding and Dosimetry. New York: Nova Science Publishers, Inc.; 2020. pp. 293-312. ISBN 9781536186611

[5] Nagaraja N, Manjunatha HC, Seenappa L, Sathish KV, Sridhar KN, Ramalingam HB. Gamma, X-ray and neutron shielding properties of boron polymers. Indian Journal of Pure & Applied Physics. 2020;**58**:271-276. DOI: 10.56042/ijpap.v58i4.67629

[6] Savaskan S. Synthesis and Investigation of Ion Exchange Properties of New Ion Exchangers [Ph.D. Thesis]. Karadeniz Technical University; 1994. Available from: https://tez.yok.gov.tr/UlusalTezMerkezi/tezSorguSonucYeni.jsp

[7] Savaskan S, Besirli N, Hazer B. Synthesis of some new cation-exchanger resins. J.Appl.Polym.Sci. 1996;**59**:1515-1524. DOI: 10.1002/(SICI)1097-4628(19960307)-59:10<1515::AID-APP3>3.0.CO;2-P

[8] Kirdsiri K, Kaewkhao J, Chanthima N, Limsuwan P. Comparative study of silicate glasses containing Bi_2O_3, PbO and BaO: Radiation shielding and optical properties. Annals of Nuclear Energy. 2011;**38**:1438-1441. DOI: 10.1016/j.anucene.2011.01.031

[9] Mortazavi SMJ, Kardan M, Sina S, Baharvand H, Sharafi N. Design and fabrication of high density borated polyethylene nanocomposites as a neutron shield. Int. J. Radiat. Res. 2016;**14**(4):379-383. DOI: 10.18869/acadpub.ijrr.14.4.379

[10] Levet A, Kavaz EE, Ozdemir Y. An experimental study on the investigation of nuclear radiation shielding characteristics in iron-boron alloys. Journal of Alloys and Compounds. 2020;**819**:152946-152955. DOI: 10.1016/j.jallcom.2019.152946

[11] Kavaz E, Gul AO, Basgoz O, Guler O, ALMisned G, Bahceci E, et al. Boron nitride nanosheet-reinforced WNiCoFeCr high-entropy alloys: The role of B4C on the structural, physical, mechanical, and radiological shielding properties. Applied Physics A. 2022;**128**:694-709. DOI: 10.1007/s00339-022-05813-5

[12] Gokmen U. Gamma and neutron shielding properties of B_4C particle reinforced Inconel 718 composites.

Nuclear Engineering and Technology. 2022;**54**:1049-1061. DOI: 10.1016/j.net.2021.09.028

[13] Lakshminarayana G, Kumar A, Tekin HO, Issa SAM, Al-Buriahi MS, Dong MG, et al. Probing of nuclear radiation attenuation and mechanical features for Lithium bismuth borate glasses with improving Bi_2O_3 content for Bi_2O_3 + Li_2O amounts. Results in Physics. 2021;**25**:104246-110459. DOI: 10.1016/j.rinp.2021-.104246

[14] Mann KS, Mann SS. Py-MLBUF: Development of an online-platform for gamma-ray shielding calculations and investigations. Annals of Nuclear Energy. 2021;**150**:107845

[15] Gerward L, Guilbert N, Jensen KB, Levring H. WinXCom-a program for calculating X-ray attenuation coefficients. Radiation Physics and Chemistry. 2004;**71**:653-654

[16] RSICC Computer Code Collection, MCNPX User's Manual Version 2.4.0. Monte Carlo N-Particle Transport Code System for Multiple and High Energy Applications; 2002

[17] Ballarini F, Battistoni G, Brugger M, Campanella M, Carboni M, Cerutti F, et al. The physics of the FLUKA code: Recent developments. Advances in Space Research. 2007;**40**:1339-1349

[18] Battistoni G, Boehlen T, Cerutti F, Chin PW, Esposito LS, Fass'o A, et al. Overview of the FLUKA code. Annals of Nuclear Energy 2015;82:10-18

[19] Sato T, Iwamoto Y, Hashimoto S, Ogawa T, Furuta T, Abe S-I, et al. Features of particle and heavy ion transport code system (PHITS) version 3.02. Journal of Nuclear Science and Technology. 2018;**55**:684-690

[20] Almuqrin AH, Sayyed MI. Gamma ray shielding properties of Yb3+-doped calcium Borotellurite glasses. Applied Sciences. 2021;**11**:5697-5701. DOI: 10.3390/app11125697

[21] Sakar E, Özpolat ÖF, Alım B, Sayyed MI, Kurudirek M. Phy-X/PSD: Development of a user friendly online software for calculation of parameters relevant to radiation shielding and dosimetry. Radiation Physics and Chemistry. 2020;**166**:108496-108508. DOI: 10.1016/j.radphyschem.2019.108496

[22] Lakshminarayana G, Elmahroug Y, Kumar A, Tekin HO, Rekik N, Dong M, et al. Detailed inspection of gamma-ray, fast and thermal neutrons shielding competence of calcium oxide or strontium oxide, comprising bismuth borate glasses. Materials. 2021;**14**:2265-2274. DOI: 10.3390/ma14092265

[23] Agostinelli S, Allison J, Amako K, Apostolakis J, Araujo H, Arce P, et al. GEANT4-A Simulation Toolkit. Nucl. Instrum. Methods Phys. Res. Sect. A Accel. Spectrometers Detect. Assoc. Equip. 2003;**506**:250-303

[24] Allison J, Amako K, Apostolakis J, Araujo H, Dubois PA, Asai M, et al. Geant4 developments and applications. IEEE Transactions on Nuclear Science. 2006;**53**:270-278

[25] Allison J, Amako K, Apostolakis J, Arce P, Asai M, Aso T, et al. Recent developments in Geant4. Nucl. Instruments Methods Phys. Res. Sect. A Accel. Spectrometers Detect. Assoc. Equip. 2016;**835**:186-225

[26] PENELOPE2014. A Code System for Monte-Carlo Simulation of Electron and Photon Transport, NEA-1525 PENELOPE2014. Available from: https://www.oecd-nea.org/tools/abstract/detail/nea-1525 [Accessed: 20 April 2020]

[27] Madbouly AM, Sallam OI, Issa SAM, Rashad M, Hamdy A, Tekin HO, et al. Experimental and FLUKA evaluation on structure and optical properties and γ-radiation shielding capacity of bismuth Borophosphate glasses. Progress in Nuclear Energy. 2022;**148**:104219-104227. DOI: 10.1016/j.pnucene.2022.104219

[28] Chanthima N, Kaewkhao J, Limsuwan P. Study of photon interactions and shielding properties of silicate glasses containing Bi2O3, BaO and PbO in the energy region of 1keV to 100GeV. Annals of Nuclear Energy. 2012;**41**:119-124. DOI: 10.1016/j.anucene.2011.10.021

[29] Al-Buriahi MS. Radiation shielding performance of a borate-based glass system doped with bismuth oxide. Radiation Physics and Chemistry. 2023;**207**:110875-110884. DOI: 10.1016/j.radphyschem.2023.110875

[30] Marzuki A, Sasmi T, Fausta DE, Harjana H, Suryanti V, Kabalci I. The effect of Bi2O3/PbO substitution on physical, optical, structural, and gamma shielding properties of Boro-tellurite glasses. Radiation Physics and Chemistry. 2023;**205**:110722-110734. DOI: 10.1016/j.radphyschem.2022.110722

[31] Mahmoud KG, Sayyed MI, Aloraini DA, Almuqrin AH, Abouhaswa AS. Impacts of praseodymium (III, IV) oxide on physical, optical, and gamma-ray shielding properties of Boro-silicate glasses. Radiation Physics and Chemistry. 2023;**207**:110836. DOI: 10.1016/j.radphyschem.2023.110836

Chapter 5

Some Features of Boron Isotopes Separation by Laser-Assisted Retardation of Condensation Method

Konstantin A. Lyakhov

Abstract

Boron isotopes have many applications in industry: medicine, semiconductor, and solar energy. Especially massive demand is for boron-10 isotopes in nuclear industry for nuclear reactors shielding and control. Various aspects of laser-assisted boron isotope separation by retardation of condensation method, such as irradiation conditions and laser and vacuum system design, have been considered. Irradiation conditions include interaction scheme of laser radiation and supersonic beam, dependence of efficiency of excitation on gas flow temperature and pressure. Basic physical constraints on laser intensity and its spectral properties have been discussed. The relation of gas flow properties, nozzle design, and vacuuming rate has been elucidated as well.

Keywords: boron isotopes, laser-assisted isotope separation, overcooled supersonic gas flow, vacuum system design, laser and optical system design, turbomolecular vacuum pump

1. Introduction

Most important applications of boron isotopes are related to boron additives widely employed in nuclear plants, as boron carbide used in control rods, or as boric acid solution used as a chemical shim in the Pressurized Water Reactors(PWR) and semiconductor industry. Due to the much larger thermal neutron absorption cross-section of boron-10 ($\sigma_n = 3837$ barn) than for boron-11 ($\sigma_n = 0.005$ barn) the use of enriched H_3BO_3 allows to reduce the total amount of boron-based poison material in the primary reactor coolant system and, therefore, to reduce corrosion and wear on the other components of the reactor core [1]. Boron-10 enriched compounds are also used to increase the efficiency of nuclear reactor emergency shutdown systems and for nuclear fuel transportation. In semiconductor industry, boron is routinely used for producing p-type domains in silica. Decreasing the size of electronic devices makes the problem of heat removal more and more important. Its solution can be using isotopically pure boron, which provides minimal distortion of the crystal structure of silica matrix and minimizes the thickness of boron acceptor layers, and, therefore, increases the heat conductivity of the acceptor layer and the transistor switching power [2]. Boron-11 can be used in this case, because protection from cosmic and other kinds of radiation becomes important with advancing miniaturization of electronic devices and in solar panels [3, 4]. Moreover, boron isotope enriched boron nitride used in nanotubes and nanoribbons has high potential for applications in nanotechnology [5–7]. As an example, in spacecraft semiconductors boron-10 enriched BN nanotubes, can be used for radiation shielding [8]. Boron-10 can be also used as a neutron-detecting component in self-powered solid-state neutron detectors [9], used in nuclear materials studies and in well logging. Boron-10 in health care is applied in boron neutron capture cancer therapy [10–12], and in studies of food properties aimed at preventing cancer and other diseases [13].

There are three methods for boron isotopes separation commonly used in industry. These methods are based on the chemical exchange reaction [14], low-temperature fractional distillation [15], and gas centrifuging. In these methods, BF_3 is used. In gas centrifuging, BCl_3 due to its high vapor pressure at room temperature, can be also used, but it is far less efficient, because of presence of three chlorine isotopes. However, using BF_3 is not economically justified, because 86% of price for high purity ^{10}B comes from powder extraction from BF_3. Hence, other methods, which rely on using BCl_3, are needed.

Laser-assisted methods comprise Atomic Vapor Laser-assisted Isotope Separation (AVLIS) [2], and Molecular Laser Isotope Separation (MLIS) methods, also known as Selection of Isotopes by Laser EXcitation (SILEX) [16–18], http://www.silex.com.au/businesses/silex. MLIS methods can be classified as following: Chemical Reaction by Isotope Selective Laser Activation (CRISLA) [19–21], Condensation Repression by Isotope Selective Laser Activation (CRISLA-2) scheme (this scheme is also known as Separation of Isotopes by Laser-Assisted Retardation of Condensation (SILARC)), and Cold Walls Laser-Assisted Selective Condensation (CWLARC) scheme (this scheme is also known as SILARC-2 [22]). Laser-assisted methods are based on the selective excitation of target isotopes in atomic or molecular form by laser radiation. Selectivity is expressed *via* specific for different target isotopes in atoms or molecules resonant-like photon energy dependence of photoabsorption cross section. In case if multiphoton dissociation is used as isotope selection mechanism, frequency of infrared (IR) laser radiation should be red-shifted [21]. Two lasers (IR + UV) can be used for dissociation or ionization of molecules. In contrast to high selectivity, subsequent chemical reactions in CRISLA method can significantly deteriorate efficiency of the process. CWLARC method has three major disadvantages: Firstly, coaxial nozzle throughout is very low, which makes it only attractive for medicine applications, secondly, wall temperature should be kept at the same low-temperature level in quite narrow interval due to the specific temperature dependence of enrichment factor, and, thirdly, isotope harvesting success strongly depends on the symmetry of selectively excited molecules. As a general advantage of SILARC methods is that only a few photons are needed for selective excitation instead of several dozens of photons required in the methods based on multiphoton dissociation (CRISLA) or ionization (AVLIS). In contrast to popular laser-assisted methods, such as Molecule Obliteration Laser Isotope Separation(MOLIS) and Atomic Vapor Laser Isotope Separation (AVLIS), large laser fluence is not required and even harmful for isotopes harvesting in SILARC methods, due to destructive resonant interaction of strong electric fields with target molecules. Ideally, laser intensity should be just high enough for molecular excitation, and, therefore, able to guide molecular dynamics in needed direction. Moreover, SILARC methods are more efficient because of typically two orders of magnitude higher linear photo-absorption cross section, than nonlinear multiphoton absorption cross section [23]. Moreover, the controllability of sequential isotope scrambling effects is more easier. The total energy consumed by vacuum pumps to provide optimal pressure level in discharge chamber ($10^{-2} - 1$ torr), is normally significantly smaller in MLIS methods than in AVLIS, where requirements for vacuum level in discharge chamber are exceptionally high ($10^{-6} - 10^{-7}$ torr). Therefore, only SILARC method deserves more detailed analysis.

SILARC method conceived by Y.T. Lee in Ref. [24], and developed by Jozef Eerkens in [22, 25]. In this method isotopes harvesting is based upon well established mass separation effect in overcooled supersonic gas flow: monomers escape gas flow core at higher rate than van der Waals clusters (dimers and higher oligomers). The larger mass difference the more separation effect is pronounced. In order to produce gas flow with uniform pressure distribution, specially profiled supersonic nozzle should be designed. Formed oligomers can be drawn from the flow either by some cold surfaces (wavy plates or walls as in SILARC-2) [17, 26], or by skimmer blade as in SILARC scheme [26]. Viability of this method was originally demonstrated on example of sulfur isotope separation in SF_6 target gas mixed with argon [27, 28].

At the first glance, the isotope production rate should increase with increasing nozzle throughput, which can be achieved by increasing nozzle dimensions, number of nozzles, or gas flow density. In the latter case, applying the laser radiation to a dense gas flow is not able to change mass distribution of formed clusters, because irreversible cluster growth sets in (in majority of clusters are over-critical [29]), and, therefore, monomer molar fraction available for excitation is rapidly decreasing. If gas flow density is not too high, then quantum optimal control methods can be helpful [30]. Too high gas density corresponds to the case when cluster loading is larger than the critical one, so-called over-critical loading. In this case, clusters are normally ionized by electromagnetic radiation but not decouple [31, 32]. However, if the gas flow dilution is high enough, its temperature is low, and transition time of molecules across irradiation cell (IC) is not too long, then the population of under-critical clusters is decreasing much slower. The upper limit for IC length is given by the condition, that the fraction of under-critical clusters in the gas flow is vanishingly small because most of them escaped gas flow. Apparently, only in the case, when under-critical clusters are still in the gas flow irradiation zone, selective resonant-like absorption of laser photons by target gas molecules can lead to retardation of their further binding with surrounding carrier gas molecules. Here, it will be considered only such gas flow pressures and temperatures that gas flow is mostly represented by monomers and dimers. We demonstrated, that increasing nozzle dimensions will decrease product cut and overall performance of isotope separation process. Therefore, only increasing the number of nozzles is the only option left to control.

Since enrichment factor and product cut are small, isotopes should be extracted by many recirculations of the gas flow. Calculations of the product cut, describing degree of gas flow separation, and enrichment factor, describing isotope content in the separated part of the gas flow, were carried out on the basis of static approximation of the transport model developed in Ref. [26].

In Ref. [33] we developed an iterative scheme for one-stage cascade and in Ref. [34] for two-stage cascade. Each cycle in this scheme corresponds to transition time of gas molecule through IC.

Configuration of the skimmer inlet should correspond to population profile of excited molecules across the separation cell cross section at the end of IC. In order to keep at minimum gas flow disturbance induced by skimmer inlet, it should be also carefully designed [35, 36].

Optimal pressure and temperature fields inside the gas flow can be provided by adequate nozzle wall shape, IC and skimmer chamber geometry, and proper choice of core and rim gas evacuation rates. Also, due to specific pattern of velocity field, yielded by the given nozzle design, the efficiency of isotope extraction is directly related to the skimmer inlet configuration and its distance from the nozzle outlet.

Since optimal enrichment facility design should be a compromise between selectivity (or time spent for extraction of a given amount of isotopes) and related energy consumptions (should be made as small as possible), we suggest that adequate criterion for efficiency of separation should correspond to the minimum of the objective function, which has a meaning of average over time of total energy consumed per unit isotope recovery [37].

In Refs. [38–40], it has been demonstrated that using an irradiation cell for isotope separation in SILARC scheme as an absorber part of CO_2 laser resonator can substantially cut electricity consumption. The idea to use a resonator as a multi-pass cavity for isotopes separation is not a novel one. For instance, as shown in Ref. [41], the use of resonator allows to increase production of selectively ionized ^{168}Yb by the order of magnitude. Moreover, in Ref. [42], it has been shown that the selectivity of dissociation of CF_2HCl molecules by CRISLA method can be significantly increased, if they are introduced inside resonator cavity of CO2 laser.

In case of overcooled gas flow, photo-absorption spectra are very narrow (only collisional broadening is dominant). Hence, the proper choice of laser pulse spectrum is crucial. In the case of CW-mode irradiation, hitting the target isotopologues is especially problematic for CO_2 laser, because it's emission spectrum is only discretely tunable, while the laser emission line should coincide with center of photo-absorption line with accuracy not worse than its full width at half of maximum (FWHM). At room temperature, the fundamental vibrational mode ν_3 of BCl_3 can be excited by CO_2 laser pulses, having emission lines 10P(8)-10P(28) in their spectrum. At laser intensities lower than saturation limit, 10P(8)-10P(16) lines are better absorbed [43], while at larger laser intensities, where multiphoton processes start to play essential role, photo-absorption maximums are more red-shifted 10P(24)-10P(28) [19, 44, 45]. By increasing pressure in the laser medium, frequency mismatches can be compensated due to emission line broadening effects. However, due to high collision rate at room temperature, the rate of excitation loss equals to or even lower than the rate of excitation gain. Hence, in order to diminish it, the gas should be sufficiently dilute and cold. These conditions both can be fulfilled by supersonically expanding rarefied gas flow.

This chapter is of the following structure. In Section 2, the transpart model to describe main features of the SILARC method is presented. In subsection 2.1, some details of evaluation of spectral properties of photo-absorption cross section are given. In subsection 2.2, definitions and results of calculation of product cut and enrichment factor are given. In Section 3, the operational principles of boron isotopes separation facility and physical processes, they are based on, are elucidated. In Section 4, gas flow irradiation conditions, such as laser spectrum choice and beam radius, are given. In subsection 4.1, basic constraints on laser intensity variation range are discussed. In Section 5, gas flow properties and requirements applied on vacuum system to provide their optimal choice are given. In subsection 5.1 examples of calculation of optimal nozzle wall shape are given. In subsection 5.2, the calculation of minimal required vacuum pump rate related to the given nozzle profile is given schematically.

2. Transport model

Excitation dynamics of the overcooled gas flow, controlled by selectively tuned laser field for this target gas, can be modeled by the transport equations, describing population dynamics of four characteristic groups of species, presented in dilute enough overcooled gas flow [26]. Molar fractions of these characteristic groups fulfill the following material balance equation:

$$f_{im} + f_{i*} + f_{i!} + f_{id} = 1, \tag{1}$$

where f_{im}, f_{i*}, $f_{i!}$, and f_{id} are molar fraction of monomers, excited monomers, epithermal, and dimers of ith isotopologue of the target gas molecule, respectively. Population dynamics of these characteristic groups are described by the following system of equations

$$\begin{cases} \dfrac{df_{i*}}{dt} = k_A f_{im} - f_{i*}\{(1-e_*)(k_{df} + k_{VV} + k_{VT} + k_{se}) + e_* k_W\}, \\ \dfrac{df_{i!}}{dt} = (1-e_*)(k_{df} + k_{VT}) f_{i*} - \{(1-e_1)k_{th} + e_1 k_{W1}\} f_{i!}, \\ \dfrac{df_{id}}{dt} = k_{df} f_{im} - k_{dd} f_{id}. \end{cases} \tag{2}$$

Here, k_{df} and k_{dd} are dimer formation and dissociation rates, k_{VT} and k_{VV} are vibration-translational and vibration-vibrational relaxation rates, k_{se} is photon spontaneous emission rate, k_{th} is epithermals thermalization rate, k_{W1} is "wall" escape rate for epithermals, e_* and e_1 are to-the-wall survival probabilities for excited monomers and epithermals, respectively. In our calculations, it was taken into account, that transport coefficients depend on the respective isotopologue mass (index i will be omitted below when no confusion occurs). In our case with sufficiently low gas flow pressure and target gas molar fraction, vibration-translational k_{VT}, vibration-vibrational k_{VV} , and photon spontaneous emission rate k_{se} are very small and can be neglected. By assuming steady gas flow and continuous wave irradiation, the system transport equations are reduced to the system of algebraic equations, which was solved in Ref. [26].

Laser excitation rate is given by:

$$k_A = \frac{\sigma_{abs}(\nu_{ex})}{h\nu_{ex}} \frac{t_{int}}{t_{tr}} \phi_L, \tag{3}$$

where $t_{int} = \frac{2R_L}{U}$ is gas flow transition time across laser beam in case of continuous wave irradiation and $t_{int} = \nu_p \tau_p t_{tr}$ in case of pulsed irradiation, and ϕ_L is the laser field distribution, which is assumed to be pencil-like with radius R_L, which corresponds to full cross-wise overlap with planar gas flow(for higher nozzle throughput):

$$\phi_L = \frac{I_{las}}{\pi R_L^2}, \tag{4}$$

where $I_{las} = I_{CW}$ in case of CW-lasing, and $I_{las} = I_p$ in the case of pulsed lasing.

2.1 Evaluation of photo-absorption cross section

According to Refs. [44, 46, 47], BCl_3 molecules photo-absorption spectra in gaseous state are regularly blue-shifted by $\sim 8-9\ \text{cm}^{-1}$ in respect to their values for BCl_3 isolated in argon matrix [48]. Since BCl_3 belongs to one of two symmetry groups, D_{3h} or C_{2v}, depending on whether the masses of chlorine isotopes are the same or not, the lines, corresponding to $\tilde{\lambda}_1$ and $\tilde{\lambda}_4$ are contributed by both, so that respective absorption peak intensities are proportional to the following probabilities[1]:

$$p^{(1)} = P\Big((35)^3\Big) + 1.5P\Big((35)^2(37)\Big) = 0.64; p^{(4)} = P\Big((37)^3\Big) + 1.5P\Big((37)^2(35)\Big) = 0.08;$$
$$p^{(2)} = 1.5P\Big((35)^2(37)\Big) = 0.21; p^{(3)} = 1.5P\Big((37)^2(35)\Big) = 0.06. \tag{5}$$

Measured distances between neighbor peaks are

$$\Delta\tilde{\lambda}_{12} = 1.41\ \text{cm}^{-1}; \Delta\tilde{\lambda}_{23} = 1.189\ \text{cm}^{-1}; \Delta\tilde{\lambda}_{34} = 1.301\ \text{cm}^{-1}. \tag{6}$$

Linear photo-absorption cross sections for each line can be extracted from Naeperian infrared intensity

$$\begin{aligned} I_{IR} &= \frac{2}{\pi}\frac{1}{27.648}\int \sigma_{abs}\big(\tilde{\lambda}, \tilde{\lambda}_k\big) d\tilde{\lambda}, \\ \sigma_{abs}\big(\tilde{\lambda}, \tilde{\lambda}_k\big) &= \sigma_{abs}\big(\tilde{\lambda}_k\big)\frac{\Delta\tilde{\lambda}}{4\big(\tilde{\lambda}-\tilde{\lambda}_k\big)^2 + \Delta\tilde{\lambda}^2}. \end{aligned} \tag{7}$$

According to experimental data for $^{11}\nu_3$ absorption band from [47]: $I_{IR} = 231.3\ \text{km/mole}$.

Since overcooled gas flow is irradiated cross-wise in opposite directions, Doppler broadenings in respect to laser beam outgoing from laser source and reflected from retro-mirror will compensate each other. Hence, photo-absorption cross section can be approximated by Lorentzian, having FWHM

$$2\Delta\nu_P = \frac{4p_{tot}\sigma_{QG}}{\pi\sqrt{\pi k_B T_{flow} M_{Q} c}} = c\Delta\tilde{\lambda} = 2.31\ \text{MHz}. \tag{8}$$

Thus, the total cross-section of isotopologues of sort k is given by:

$$\begin{aligned} \sigma_{abs}^{(k)} &= \sigma_{\max}^{(k)}\frac{\Delta\nu_P^2}{(\nu-\nu_k)^2 + \Delta\nu_P^2}, \\ \sigma_{\max}^{(k)} &= \frac{27.648 I_{IR}}{\Delta\tilde{\lambda}} p^{(k)}, \end{aligned} \tag{9}$$

where

$$\Big\{\sigma_{\max}^{(1)}, \sigma_{\max}^{(2)}, \sigma_{\max}^{(3)}, \sigma_{\max}^{(4)}\Big\} = \{17.72; 5.74; 1.84; 2.23\} \times 10^{-16}\ \text{m}^2. \tag{10}$$

---

[1] probability to find combination of $^{35}Cl_n$ and $^{37}Cl_{3-n}$ is denoted by $P\Big((35)^n(37)^{3-n}\Big)$.

2.2 Product cut and enrichment factor evaluation

Product cut can be evaluated as a fraction of escaped target gas molecules and oligomers from the gas flow core

$$\theta = \frac{Q_{\text{esc}}}{Q_{\text{feed}}}, \tag{11}$$

where $Q_{\text{feed}} = y_Q N_{\text{feed}}, N_{\text{feed}} = Q_f t_{\text{tr}}$ is the number of target gas molecules in the feeding gas flow and Q_{esc} is the number of escaped specied. The product cut can be transformed to $\theta = \mathfrak{a} + \mathfrak{b}$, where

$$\begin{aligned} \mathfrak{a} &= \sum_{\substack{k=1 \\ k \neq i}}^{2} x_k \alpha_k, \quad \alpha_k = (1 - f_{kd})\Theta + f_{kd}\Theta_d; \\ \mathfrak{b} &= x_i \alpha_i, \quad \alpha_i = (1 - f_{i!} - f_{id})\Theta + f_{i!}\Theta_1 + f_{id}\Theta_d, \end{aligned} \tag{12}$$

relative isotope abundance of excited ith isotope is denoted as x (for definiteness we assume that only $^{11}BCl_3$ isotopologues are excited, i.e., $i = 2$), and

$$\Theta = 1 - e^{-k_W t_{\text{tr}}} \tag{13}$$

is a fraction of escaped from gas flow core monomers,

$$\Theta_1 = 1 - e^{-\mu_1 k_W t_{\text{tr}}} \tag{14}$$

is a fraction of escaped epithermals, and

$$\Theta_d = 1 - e^{-\psi_d k_W t_{\text{tr}}} \tag{15}$$

is fraction of escaped dimers, where

$$\psi_d = \frac{\sigma_{cQ/G} M_{Q/G}^{1/2}}{\sigma_{cQG/G} M_{QG/G}^{1/2}}. \tag{16}$$

Enrichment factor by ith isotope is given by

$$\beta = \frac{{}^i Q_{\text{esc}}/Q_{\text{esc}}}{{}^i Q_{\text{feed}}/Q_{\text{feed}}} = \frac{\alpha_i}{\theta}, \tag{17}$$

where ${}^i Q_{\text{feed}}/Q_{\text{feed}}$ and ${}^i Q_{\text{esc}}/Q_{\text{esc}}$ are relative abundances of ith isotope in the escaped and feed gas flows, respectively.

Product cut and enrichment factor as functions of gas flow core temperature (other parameters are fixed to their optimal values) were calculated in Refs. [37, 49]. Significant descrepancy between the maximal values of enrichment factors 1.0043 ($p_{tot} = 10$ mTorr, $T = 24$ K, $\sigma_{abs} = 13.75$ Mb, $\phi_L = 39.31$ W/cm^2, $L_{IC} = 67$ cm) and 1.25 ($p_{tot} = 10$ mTorr, $T = 25$ K, $\sigma_{abs} = 0.071$ Mb, $\phi_L = 10$ kW/cm^2, $L_{IC} = 20$ cm), obtained in [37, 49] respectively, can be explained by that the formula (32) from [26] should be multiplied by the vibration-translational(VT) transition probability, as it should be in order to get proper equilibrium dimer concentration (Eq. (124) in [25]). Here, L_{IC} is gas flow length from nozzle outlet to skimmer inlet.

3. Operational principles

At the beginning of operation, the mixing tank is occupied only by carrier gas, while target gas molecules, having natural relative isotope abundance, are stored in the feed chamber. Then, the target gas is injected into the mixing tank. Target gas is seeded at a very low molar fraction($y_Q \sim 0.02$) into carrier gas, in order to minimize nearly resonant VV excitation loss caused by collisions among BCl_3 molecules. Gas flow should be diluted as well in order to minimize nearly resonant excitation loss due to VT (vibration-translational) relaxation. In order to provide pressure $p_{flow} \approx 10$ mtorr and temperature $T_{flow} \approx 24$ K, that correspond to maximum of enrichment factor, provided Ar is chosen as a carrier gas, the total pressure in the mixing tank should be $P_0 \approx 6$ torr [38], provided expansion remains isentropic at least along the gas flow core axis.

The scheme shown in **Figure 1** allows multifrequency boron isotopes excitation in the absorber part of CO_2 laser resonator (in Ref. [39] the same scheme was proposed, but for orthogonal gas flow irradiation). From the mixing tank gas expands

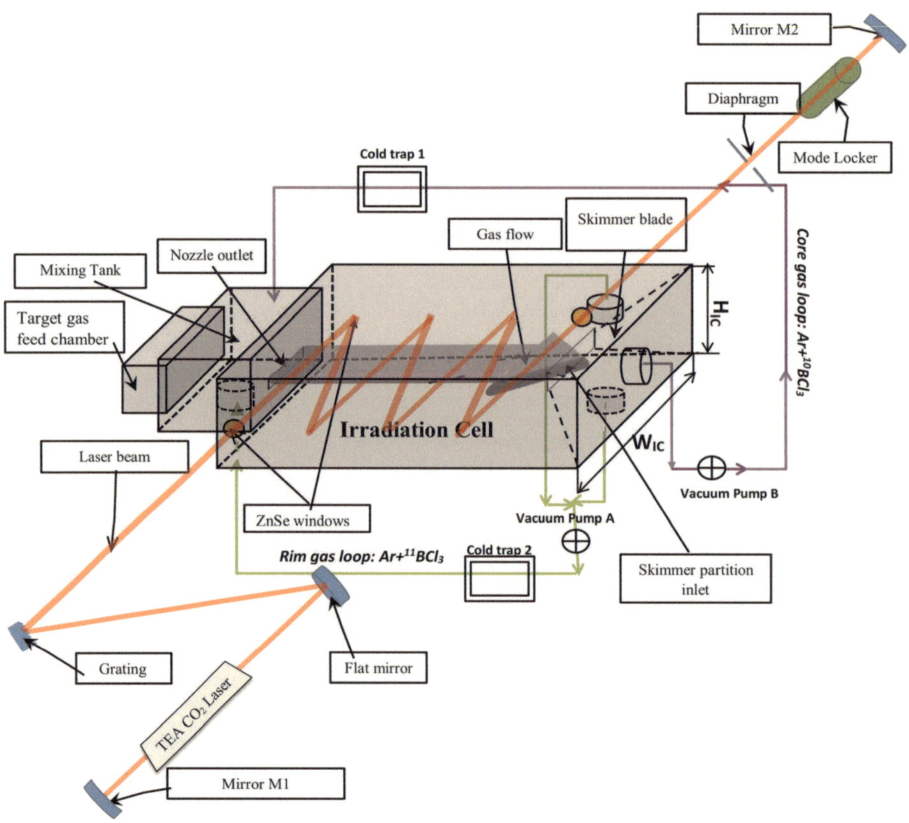

Figure 1.
The principal scheme of boron isotope enrichment setup (edges, links, and details, that are invisible from the frontal surface, are displayed by dashed lines). The laser beam passes through a semi-transparent window and cross-wise impinges the gas flow. In order to increase excitation rate, the laser beam oscillates within a resonant multi-pass cavity between opposite mirror strips in order to interact with whole length of supersonic jet. Core gas is boron-10 enriched by selective laser evaporation of $^{11}BCl_3$ isotopologues. It is circulated from the mixing tank and back in order to compensate pressure loss and to achieve a higher degree of target isotope recovery.

continuously into the separation chamber, where it is irradiated cross-wise by the laser beam confined between mirror strips placed on opposite walls of resonator, three (one flat and two concave) mirrors outside the resonator cavity and grating.

Multiple reflections from mirrors may lead, however, to the instability of laser power and frequency[2]. As shown in Refs. [38, 39], the use of resonant absorber cavity for isotope separation will lead to 4–6.5 times increase of laser pulse peak intensity.

In order to provide the largest gas flow irradiation volume, having the optimal static pressure and temperature distributions, a uniform velocity profile can be provided by slit nozzle design, which is discussed in more detail in subsection 5.1.

Skimmer blade divides the gas flow into target isotope (boron-11) enriched (rim) and desired isotope (boron-10) enriched (core) fractions by the end of chamber. Core and rim gas flow fractions, having laser-controlled target isotopologue populations, are evacuated separately by two separate vacuum pumps, having appropriate pumping down rates. Before redirecting them back into mixing tank or exhausting into atmosphere, they are captured in respective Zeolite cold traps as shown in **Figure 1**. Roughly speaking, pumping out speeds should be chosen proportional to corresponding stagnation pressures. The larger the initial pressure loss, the less demanding requirements for vacuum pumps. Material, momentum, and energy balance equations for skimmer partition have been solved in Section 5.2. Besides evaluating required minimal levels for pumping down rates for core and rim gas flows, it helps to clarify contributions and relations between various parameters, affecting gas flow deceleration, and, therefore, stagnation pressures, corresponding to each vacuum pump.

4. Irradiation conditions

According to experimental data, the gaseous state $^{11}BCl_3$ photo absorption line center position ranges from $\tilde{\lambda}^{(1)}_{\text{max}} = 957.67\ \text{cm}^{-1}$, as of [44], to $\tilde{\lambda}^{(1)}_{\text{min}} = 954.2\ \text{cm}^{-1}$, as of [47], according to different physical conditions (more red-shifted spectra correspond to higher gas temperatures). The value $944.194\ \text{cm}^{-1}$, measured in Ref. [19], corresponds to multiphoton absorption.

If photoabsorption takes place at physical conditions, corresponding to $\tilde{\lambda}^{(1)}_{\text{max}}$, then FWHM, corresponding to 10P(4)-10P(10) emission lines, should be:

$$\Delta\tilde{\lambda}^{(k)}_{\text{FWHM}} = \{-0.26; 0.142; 1.05; 1.77\}\ \text{cm}^{-1}, \tag{18}$$

where 10P4 line (1-st entry) is almost vanishing, and line 10P6(2-nd entry) is very weak, so only line 10P8 can be used for one ($\Delta\tilde{\lambda}^{\text{max}}_{\text{FWHM}} = 1.05\ \text{cm}^{-1}$) or two ($\Delta\tilde{\lambda}^{\text{max}}_{\text{FWHM}} = 1.31\ \text{cm}^{-1}$) isotopologues excitation. In case, if photoabsorption takes place at physical conditions, corresponding to $\tilde{\lambda}^{(1)}_{\text{min}}$, then FWHM, corresponding to 10P(8)-10P(14) emission lines, should be:

[2] As a possible ways to overcome this problem, adaptive retro-mirror surface automatic adjustment or its displacement by piezoelectric transducer (PZT), which is activated by generator, that is controlled by phase-shift sensitive detector.

$$\Delta\tilde{\lambda}_{\text{FWHM}}^{(k)} = \{-0.69; -0.19; 0.81; 1.632\} \ \text{cm}^{-1}. \tag{19}$$

10P(8)-10P(12) lines can be used for simultaneous excitation of all three most abundant chlorine isotopologues, provided laser medium pressure is 5.75 bars, which corresponds to $\Delta\tilde{\lambda}_{\text{FWHM}}^{\max} = 0.81 \ \text{cm}^{-1}$ and laser pulse width $\tau_p = 334 \ \text{ps}$.

4.1 Admissible laser intensity variation range

Absorbed energy, corresponding to pulsed excitation of chlorine isotopologues of $^{11}BCl_3$, can be estimated from the formula for excitation rate, deduced in Ref. [39]:

$$k_A(t) = \frac{1}{2\pi^2 R_L^2} \sum_{k=1}^{N_l} \int_{-\infty}^{\infty} \frac{d\omega}{\hbar\omega} e^{I\omega t} \frac{dE_A^{(k)}}{d\omega}, \tag{20}$$

where it was assumed, that laser beam is pencil-like ($R_L = 2.8 \ \text{mm}$ is laser beam radius, corresponding to the thickness of gas flow core with optimal pressure and temperature inside). It can be represented as:

$$k_A(t) = \frac{1}{\pi R_L^2} \sum_{i=1}^{N_l} \sum_{j=1}^{N_l} F^{-1}\left[\frac{d\lambdabar^{(ij)}}{d\omega}\right], \tag{21}$$

where

$$\frac{d\lambdabar^{(ij)}}{d\omega} = \frac{1}{\hbar\omega} \frac{dE_A^{(ij)}}{d\omega} \tag{22}$$

is absorbed photons spectral density. As shown in Ref. [39], laser pulse spectral density can be represented as:

$$\frac{dE_A^{(ij)}}{d\omega} = E_0^{(i)} E_0^{(j)} \ {}^{(i)}(t) \sigma^{(j)}(\omega) \daleth^{(j)}(\omega). \tag{23}$$

Let us see what is the average number of resonantly absorbed photons per BCl_3 molecule, provided laser pulse intensity I_p^{ref} corresponds to some reference CW-mode intensity $I_{CW}^{ref} = 1\text{W}$ and BCl_3 absorption spectrum, corresponding to $\tilde{\lambda}_{\min}^{(1)}$, was chosen. According to the formula, derived in Ref. [39], average numbers of photons in pulsed mode, absorbed by each isotopologue excitable by different lines, are:

$$\left\{N_{\text{ph}}^{(1)}, N_{\text{ph}}^{(2)}, N_{\text{ph}}^{(3)}, N_{\text{ph}}^{(4)}\right\} = \{4.5, 77.8, 0.26, 0.57\} \times 10^{-3}. \tag{24}$$

It is seen, that majority of photons are absorbed by ν_2 line, which is in the closest proximity to 10P8. However, the number of photons, absorbed within the excitation loss characteristic time interval, should be less than $N_{crit}^{max} = 34 - 39$ [50], in order to avoid multiphoton dissociation. Excitation loss time coincides with gas flow transition time across laser beam, because all other relaxation times, such as dimer formation, vibration-vibrational, and vibration-translational transfers as well as spontaneous emission, are bigger [38]. Thus, the upper limit for laser intensity should be

$$I_{\max} = I_{\text{ref}} \frac{N_{crit}^{max}}{Max\left\{N_{\text{ph}}^{(k)}\right\}} = 437 - 501 \text{ W}. \tag{25}$$

The lower limit for applicability of SILARC method is, apparently, given by the following condition:

$$I_{\min} = \frac{I_{\text{ref}}}{Max\left\{N_{\text{ph}}^{(k)}\right\}} = 12.8 \text{ W}. \tag{26}$$

In the case, when BCl_3 absorption spectrum, corresponding to $\tilde{\lambda}_{\max}^{(1)}$, was chosen, the average number of photons, absorbed by each absorption line, from 10P8 emission line are:

$$\left\{N_{\text{ph}}^{(1)}, N_{\text{ph}}^{(2)}, N_{\text{ph}}^{(3)}, N_{\text{ph}}^{(4)}\right\} = \{0.02, 0.1, 0.41, 0.29\} \times 10^{-3}. \tag{27}$$

Thus, the upper and lower limits for laser intensity should be $I_{\max} = 83.2 - 94.4\text{kW}, I_{\min} = 2.45 \text{ kW}$.

5. Gas flow properties and vacuum system design

Apparently, the main goal in vacuum system design is to provide the largest overlap of laser beam with gas flow core, having optimal pressure and temperature for isotope separation. This condition in the case of cross-wise irradiation can be fulfilled, provided gas flow is planar (interaction region with cross-wise directed laser beam is large), and preserves its shape over all its extension. This can be implemented, if the gas flow is perfectly expanded. Thus, the vacuum chamber pumping down speed should be carefully chosen. The pumping out speed is provided by two vacuum pumps. High-rate one evacuates the central part of the gas flow, and low-rate one evacuates the peripheral gas flow. The main gas flow stream is divided into two parts by the wedge-like inlet of the skimmer chamber, which is placed on the opposite side of the irradiation cell.

5.1 Nozzle design

Nozzle wall shape was found as a friction-free(isentropic) region corrected by the boundary layer. According to experimental data from Ref. [51], gas flow deceleration (full pressure drop) across the nozzle practically does not depend on angle of nozzle inlet, if $\theta_0 < 85^0$. The larger radius of the nozzle throat curvature R_2, the larger gas flow uniformity, and, therefore, the smaller deceleration. According to boundary layer theory, in order to take into account dissipative effects, nozzle contour, corresponding to isentropic gas flow expansion, should be corrected by displacement layer thickness (DLT).

In order to figure out which gas flow regime takes place-laminar or turbulent, one needs to know Reynolds number evolution along the nozzle axis. Since boundary

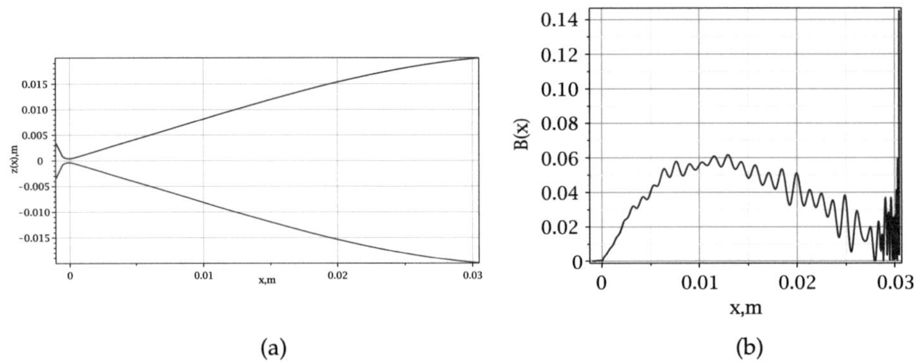

Figure 2.
(a) Cross-section of nozzle profile in zx plane (gas flow temperature at the nozzle outlet along its axis (it's minimal value) corresponds to optimal value for boron isotopes separation) and (b) break down parameter distribution along the nozzle axis for boron isotopes separation.

layers from opposite walls do not interfere with each other, Reynolds number to characterize the transition from laminar to turbulent flow can be introduced as:

$$Re_x = \frac{\rho v x}{\mu}. \tag{28}$$

If $Re_x > 5 \times 10^5$, then gas flow gets certainly turbulent for most commercial surfaces. Therefore, for our nozzle design, it should be definitely stable, since $Re_x(X_{fin}) = 2615$.

The boundary layer corrected nozzle profile for argon used as a carrier gas and break down parameter is shown in **Figure 2a**.

Calculations of continuous gas flow are only valid until breakdown parameter $B = M\sqrt{\frac{\pi\gamma}{8}}\frac{\lambda}{\rho}\left|\frac{d\rho}{dz}\right|$ is lower than its critical value $B_{crit} = 0.05$. As seen in **Figure 2b**, gas flow departs from continuous regime practically overall nozzle extension. Therefore, caution must be exercised relative to obtained results on nozzle profile.

5.2 The model for estimation of minimal requirements for vacuuming rate

Influence of gas flow deceleration, caused by interaction with ambient gas and irradiation chamber-vacuum pump inlet tract, on temperature distribution in gas flow can be tolerated, if pressure builds up at vacuum pump inlet still provides an acceptable variation of static pressure for isotope separation in the discharge chamber. This condition can be fulfilled if vacuum tract resistance to the gas flow is reduced to a minimum and pumping out speed is accurately chosen. As well known, during gas flow transition across vacuum tract, gas pressure rises due to friction with walls, or, in the case of gas flow injected in the pipe of much larger diameter, with ambient gas, due to shock waves caused by hydrodynamic discontinuities.

Let us consider subsequent stages of gas flow evolution. In the beginning, gas chamber is pumped down until some pressure level, which is smaller than the pressure at nozzle outlet. In this case, gas flow gets under-expanded, and occupies the space between irradiation chamber walls, according to pressure of residual gas. On the next stage, ambient gas pressure increases accordingly to pumping down rate applied, so that angle between gas flow direction and oblique shock at nozzle outlet decreases.

Since we need perfectly expanded gas flow, vacuum pump characteristics should be such that this angle vanishes after some time and then does not change.

As pointed out in Ref. [36], the large decrease in stagnation pressure across a normal shock wave may be reduced by decelerating the free-stream flow by means of one or more oblique shock waves, followed by a weak normal shock wave, produced by central body, placed at skimmer inlet. By employing that principle, an efficient external compression of the gas flow is achieved before the gas flows into a subsonic internal compression diffuser (skimmer partition of discharge chamber). Larger stagnation pressure recoveries are obtainable if several successive weak shock waves instead of one relatively strong conical shock wave are utilized for decelerating the gas flow.

To find gas density, one needs to solve a system of material, momentum, and energy conservation equations. For sake of simplicity, this system of equations can be transformed, according to hydraulic approximation, to the system of balance equations, where taking an average of conservation equations is carried out over cross sections at the characteristic intermediate stations i: The first station$(i = 1)$ corresponds to inlets of the pipes, connected to the port of rim or core gas flow evacuating vacuum pumps, the second one$(i = 2)$ corresponds to core gas flow evacuating vacuum pump inlet or to two pipe into one pipe connection, assigned for rim gas flow evacuation, the third one$(i = 3)$ corresponds to rim gas flow evacuating vacuum pump inlet. This system of balance equations is following:

$$\begin{cases} \dfrac{d\mathcal{N}^{a;i}}{dt} + \rho_{in}^{a;i} v_{in}^{a;i} A_{in}^{a;i} = \rho_{out}^{a;i} v_{out}^{a;i} A_{out}^{a;i}, \\ \dfrac{d\mathcal{P}^{a;i}}{dt} + A_{in}^{a;i}\left(P_{in}^{a;i} + m_0 \rho_{in}^{a;i}\left[v_{in}^{a;i}\right]^2\right) + \delta P^{a;i} = A_{out}^{a;i}\left(P_{out}^{a;i} + m_0 \rho_{out}^{a;i}\left[v_{out}^{a;i}\right]^2\right), \\ \dfrac{d\mathcal{E}^{a;i}}{dt} + \rho_{in}^{a;i} v_{in}^{a;i} A_{in}^{a;i}\left(\dfrac{m_0\left[v_{in}^{a;i}\right]^2}{2} + C_p T_{in}^{a;i}\right) + \delta Q^{a;i} = \rho_{out}^{a;i} v_{out}^{a;i} A_{out}^{a;i}\left(\dfrac{m_0\left[v_{out}^{a;i}\right]^2}{2} + C_p T_{out}^{a;i}\right), \\ P_{out}^{a;i} = \rho_{out}^{a;i} k_B T_{out}^{a;i}. \end{cases} \tag{29}$$

In order to close this system of equations, it should be supplemented by the ideal gas equation of state

$$P_a^{out;i} = \rho_a^{out;i} k_B T_a^{out;i}, \tag{30}$$

and by equation for friction coefficient f_a^i. We assume that gas friction with pipe wall surface is laminar

$$Re_D^{a;i} = \frac{\rho_a^{out;i} v_a^{out;i} D_i}{\mu} < 2300, \tag{31}$$

where D_i is pipe diameter. Hence, the friction coefficient can be estimated as:

$$f_a^i = \frac{64}{Re_D^{a;i}}. \tag{32}$$

The total number of molecules in rim and core gas flow fractions, accumulated over its transition time across the irradiation chamber are:

$$Q_{core} = (1 - \mu\theta)Q_f \tag{33}$$

and

$$Q_{rim} = \mu\theta Q_f \tag{34}$$

respectively, where $\mu = 0.02$ is target gas(BCl_3) molar fraction. Product cut, or fraction of molecules escaped the gas flow core, was calculated on the basis of the transport model, developed in Ref. [26]. It's value, corresponding to optimal conditions $T_a^{in;0} = 25$ K, $P_a^{in;0} = 10$ mtorr, calculated in Ref. [38], is $\theta \approx 0.2$.

Heat transfer intensity can be phenomenologically approximated as:

$$q = h|T_w - T|, \tag{35}$$

where wall temperature is fixed at $T_w = T_{atm} = 300$ K, and T is average temperature over gas flow volume. A coefficient of proportionality h is introduced as $h = \rho v C_p N_{St}$, where Stanton number N_{St} is related to momentum transfer due to friction by Reynolds analogy, which, however, yields satisfactory results only for gases:

$$N_{St} = \frac{1}{2} Prf \tag{36}$$

Friction factor f is introduced as:

$$f = \frac{wall\ shear\ stress}{\frac{1}{2}\rho v^2}. \tag{37}$$

The total heat transfer can be approximated as:

$$\delta Q_a^i = \frac{\overline{f}_a^i}{2} Pr(T)\rho_a v_a C_p |T_w - T| S_a. \tag{38}$$

Average values are approximated as $\overline{o} = \frac{o_{in} + o_{out}}{2}$. We assume, that momentum transfer to the wall surface can be approximated by the formula for gas flow pressure build-up due to friction over planar surface:

$$\delta P_a^i = \frac{f_a^i}{8} m_0 \overline{\rho}_a \overline{v}_a^{\,2} S_a \tag{39}$$

Since the efficiency of pumping down speed depends on the gas density at the vacuum pump inlet, one needs to know which model of vacuum pump is able to provide the required pumping out speed. Then, given that what gas is pumped down and provided it's hold at room temperature, this dependence can be transformed into the following relation to inlet density:

$$\frac{dU^a}{dt} = \kappa f_{pump}(\rho^a), \tag{40}$$

where factor κ fits reference vacuum pump interpolated dependence on gas density to desired pumping down speed at a given density. We have used pressure dependence of pumping speed of the turbo-molecular pump (TMP) TURBOVAC MAG 3200C [52], as a reference, which is shown in **Figure 3a**, where different gases at room temperature are pumped down. Thus, the performance curve for pumping out of the air can be taken from interpolation:

$$f_{pump} = 3.07 - 3.09 \times 10^{-21}\rho(t) + 9.5 \times 10^{-43}\rho(t)^2. \quad (41)$$

Moreover, it can be estimated pressure evolution within the pumped-down vessel. According to the Reynolds transport theorem, particle balance condition leads to the following equation for average gas density ρ_a in the inter-chamber connector:

$$\begin{cases} V_{ch}\dfrac{d\rho_a(t)}{dt} = \dfrac{dN_a}{dt} - \rho_a(t)\dfrac{dU_a}{dt}, \\ \rho_a(0) = \rho_{in}, \\ \rho_{in} = \dfrac{p_{in}}{k_B T_0}. \end{cases} \quad (42)$$

The vacuum pump should provide such pumping speed, that asymptotic gas density $\rho_{as} = \lim\limits_{t\to\infty}\rho(t)$ is within the range:

$$\rho_{out} < \rho_{as} < \rho_{in}, \quad (43)$$

where T_0 is room temperature.

Solution of the balance equation at the initial condition, that stagnant gas resides in the feed chamber at room temperature and pressure 10 mtorr, for core gas flow is shown in **Figure 3b** and for rim gas in **Figure 3c**. It is seen from this figure, that after rather short period of time stationary condition is reached (it depends on the volume of intermediate chamber). Pumping speeds were chosen from the condition that asymptotic density is ranged as (43).

Inter-chamber connector geometry and pipe length and diameter for core and rim gas flows respectively should be chosen so, that separated species are not driven due to high pressure back into irradiation chamber: $P^a_{out} < P_{in}$. We have the following solutions of the system of Eqs. (29) $P^{core}_{out} = 0.27$ Pa, $T^{core}_{out} = 187.7$ K, $v^{core}_{out} = 44.68$ m/s, $\rho^{core}_{out} = 3.34 \times 10^{19}$ m^{-3}, $P^{rim}_{out} = 1.95$ Pa, $T^{rim}_{out} = 96.43$ K, $v^{rim}_{out} = 11.32$ m/s, $\rho^{rim}_{out} = 7.49 \times 10^{20}$ m^{-3}, corresponding to given geometry for core and

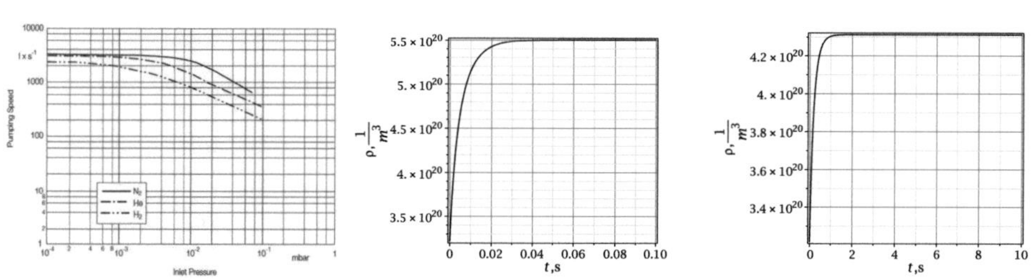

(a) (b) (c)

Figure 3.
(a) TURBOVAC MAG 3200C pumping speed as a function of inlet gas pressure, [52], (b) Core gas flow density evolution, provided TMP-3403LMC with flange VG300 is used, [53], and (c) rim gas flow density evolution, provided HiPace 300 DN100 is used.

rim gas flows. For core and rim gas flows one obtains $Re_D^{core;1} = 69$ and $Re_D^{rim,1} = 37$, respectively. Thus, $f_{rim}^1 = 0.3$ and $f_{core}^1 = 0.2$.

Thus, we obtain the following values for pumping down speeds for core and rim gas flows $\frac{dU_{core}}{dt} = 3300\ \text{l/s}$, $\frac{dU_{rim}}{dt} = 310\ \text{l/s}$.

Since the system of Eqs. (29) along with approximation for heat transfer (38), momentum loss (39), and friction coefficient (37) is valid only if gas flow is continuous along the wall surface and shock-free, while in real situations gas flow is rather injected in the pipe due to its significantly smaller cross section, comparing to the pipe area, especially in the case of rim gas flow, found pipe geometry is expected to be just a lower limit (pipe length can be larger), the same concerns to found pumping speeds for rim and core gas flows evacuation, which in turn can serve as an upper limit for pumping speed requirements.

6. Conclusions

In this chapter, physical constraints such as gas flow core temperature and pressure distributions, oblique shocks configuration in the course of gas evacuation from discharge chamber, condensation rate, pressure in CO_2 laser medium, and spectral properties of BCl_3 at different temperatures and pressures on the efficiency of SILARC method have been discussed. In particular, it has demonstrated the possiblity to extract boron isotopes from BCl_3 gas flow more efficiently by applying a three-mirros scheme for multiline excitation. Recommendations on laser system design, optimal nozzle design, and parameters of the corresponding vacuuming system are given.

Author details

Konstantin A. Lyakhov
Steklov Mathematical Institute, Moscow, Russia

*Address all correspondence to: lyakhov2000@yahoo.com

References

[1] Lamarsh JR, Baratta AJ. Introduction to Nuclear Engineering. 3rd ed. Upper Saddle River, New Jersey: Prenice Hall; 2001

[2] Bokhan PA, Buchanov VV, Fateev NV, Kalugin MM, Kazaryan MA, Prokhorov AM, et al. Laser Isotope Separation in Atomic Vapor. Weinheim: WILEY-VCH Verlag GmbH & Co. KGaA; 2006

[3] Steigerwald A et al. Determination of optical damage cross-sections and volumes surrounding ion bombardment tracks in GaAs using coherent acoustic phonon spectroscopy. Journal of Applied Physics. 2012;**112**:013514

[4] Wang PF et al. Nitrogen-promoted formation of graphite-like aggregations in SiC during neutron irradiation. Journal of Applied Physics. 2012;**111**:063517

[5] Sun C et al. Recent development of the synthesis and engineering applications of one-dimensional boron nitride nanomaterials. Journal of Nanomaterials. 2010;**2010**:1-16

[6] Han WQ et al. Isotope effect on band gap and radiative transitions properties of boron nitride nanotubes. Nano Letters. 2008;**8**(2):491-494

[7] Yu J et al. Narrowed bandgaps and stronger excitonic effects from small boron nitride nanotubes. Chemical Physics Letters. 2009;**476**(4–6):240-243

[8] Yu J et al. Isotopically enriched ^{10}BN nanotubes. Advanced Materials. 2006; **18**(16):2157-2160

[9] Dahal R et al. Self-powered micro-structured solid state neutron detector with very low leakage current and high efficiency. Applied Physics Letters. 2012; **100**:243507

[10] Adams DM, Ji W, Barth RF. Comparative in vitro evaluation of dequalinium B, a new boron carrier for neutron capture therapy (nct). Anticancer Research. 2000;**20**(5B): 3395-3402

[11] Kiger WS et al. A pharmacokinetic model for the concentration of B-10 in blood after boronophenylalanine-fructose administration in humans. Radiation Research. 2001;**155**(4): 611-618

[12] Barth RF. Boron neutron capture therapy at the crossroads: Challenges and opportunities. Applied Radiation and Isotopes. 2009;**67**:S3-S6

[13] Naghii MR, Samman S. The role of boron in nutrition and metabolism, Progress in food nutrition. Science. 1993; **4**:331-349

[14] Ye DY, Zhou YS. Review on production and separation of isotope boron 10. Fine Chemistry. 1998;**1**:37-39

[15] Huang Y, Cheng S, Jiao X, Zhang W. Research on chemical exchange process of boron isotope separation. Procedia Engineering. 2011;**18**:151-156

[16] Lyman JL. Enrichment separative capacity for SILEX, Tech. Rep. LA-UR-05-3786, Los Alamos National Laboratory, Los Alamos. 2005

[17] Eerkens JW. Separation of isotopes by laser-assited retardation of condensation (SILARC). Laser and Particles Beams. 1998;**16**(2):295-316

[18] Mathi P, Parthasarathy V, Nayak AK, Mittal JP, Sarkar SK. Laser isotope separation: Science and technology. Proceedings of the National Academy of Sciences India Section A:

Physical Sciences. 2015:1-16. DOI: 10.1007/s40010-015-0249-6

[19] Lyman JL, Rockwood SD. Enrichment of boron, carbon, and silicon isotopes by multiple photon absorption of 10.6μm laser radiation. Journal of Applied Physics. 1976;**47**:595. DOI: 10.1063/1.322619

[20] Parvin P et al. Molecular laser isotope separation versus atomic vapor laser isotope separation. Progress in Nuclear Energy. 2004;**44**:331-345

[21] Makarov GN. Selective IR excitation and dissociation of molecules in gas dynamically cooled jets and flows. Physics-Uspekhi. 2005;**48**:37-77

[22] Eerkens JW. Isotope separation by condensation reduction of laser-excited molecules in wall-cooled subsonic gas streams. Nuclear Science and Engineering. 2005;**150**:1-26

[23] Grant ER, et al. Molecular Beam Studies on Multiphoton Dissociation of Polyatomic Molecules. In: Eberly JH, Lambropoulos P, editors. Proceedings of the International Conference on Multiphoton Processes. New York: J. Wiley and Sons; 1978. pp. 359-369

[24] Lee YT. Isotope separation by photodissociation of van der wall's molecules, U.S. Patent No. 4,032,306. 28 June 1977

[25] Eerkens JW. Equilibrium dimer concentration in gases and gas mixtures. Chemical Physics. 2001;**269**:189-241. DOI: 10.1016/S0301-0104(01) 00346-9

[26] Eerkens JW. Laser-induced migration and isotope separation of epithermal monomers and dimers in supercooled free jets. Laser and Particle Beams. 2005;**23**:225-253. DOI: 10.1017/S026303460522325X

[27] Zellweger JM. Isotopically selective condensation and infrared-laser-assisted gas-dynamic isotope separation. Physical Review Letters. 1984;**52**(7):522-525

[28] den Bergh V. Laser assisted aerodynamic isotope separation. Laser und Optoelectronik. 1985;**3**:263

[29] Eerkens JW. Nucleation and particle growth in super-cooled flows of SF_6/N_2 and UF_6/N_2 mixtures. Chemical Physics. 2003;**293**:112-153. DOI: 10.1016/S0301-0104(03)00300-8

[30] Koch C et al. Quantum optimal control in quantum technologies. Strategic report on current status, visions and goals for research in europe. EPJ Quantum Technology. 2022;**9**:1-60

[31] Blasko F et al. Characterization of argon cluster jets for laser interacion studies. Nuclear Instruments and Methods in Physics Research B. 2003; **205**:324-328

[32] Arefiev AV. Size distribution and mass fraction of microclusters in laser-irradiated plasmas. High Energy Density Physics. 2010;**6**:121-127

[33] Lyakhov KA, Pechen AN. Constrained optimization criterion for zirconium isotopes separation by the laser assisted retardation of condensation method. Proceedings of the Steklov Institute of Mathematics. 2021;**313**:1-11

[34] Lyakhov KA. Optimization criterion for two-stage Cascade for iterative molybdenum isotopes recovery by the method of laser-assisted retardation of condensation(SILARC). Journal of Physics: Conference Series. 2022;**2147**: 012009. DOI: 10.1088/1742-6596/2147/1/012009

[35] Abramovich GN. Applied Gas Dynamics. 3rd ed. Wright-Patterson Air Force Base, Ohio: Foreign Technology Division; 1973

[36] Oswatitsch K. Pressure recovery for missiles with reaction propulsion at high supersonic speeds (the efficiency of shock diffusers), Tech. Rep. NACA-TM-1140, National Advisory Committee for Aeronautics, Washington, DC United States. 1947

[37] Lyakhov KA, Lee HJ, Pechen AN. Some issues of industrial scale boron isotopes separation by the laser assisted retarded condensation(silarc) method. Separation and Purification Technology. 2017;**176**:402-411

[38] Lyakhov KA, Lee HJ, Pechen AN. Some features of boron isotopes separation by the laser-assisted retardation of condensation method in multipass irradiation cell implemented as a resonator. IEEE Journal of Quantum Electronics. 2016;**52**:1400208–1-1400208–8

[39] Lyakhov KA, Pechen AN. CO_2 laser system design for efficient boron isotope separation by the method of selective laser-assisted retardation of condensation. Applied Physics B: Lasers and Optics. 2020;**126**:141–1–141-11

[40] Lyakhov KA, Pechen AN. Laser and diffusion driven optimal discrimination of similar quantum systems in resonator. Lobachevskii Journal of Mathematics. 2022;**43**(7):1693-1703. DOI: 10.1134/S1995080222100249

[41] Tkachev AN, Yakovlenko SI. Use of cavity as a multipass system in laser isotope separation. Quantum Electronics. 1997;**24**(8):759-762

[42] Lokhman VN, Makarov GN, Ryabov EA, Sotnikov MV. Carbon isotope separation by infrared multiphoton dissociation of CF_2HCl molecules with a separation reactor in a laser cavity. Quantum Electronics. 1996; **26**:79-86

[43] Lavigne P, Lachambre JL. Pressure dependent absorption in BCl_3 at 10.6 μm. Applied Physics Letters. 1971; **19**:176-179

[44] Karlov NV, Petrov YN, Prokhorov AM, Stel'mach OM. Dissociation of boron trichloride molecles by CO_2 Laser radiation. Letters to the Soviet Journal of Experimental and Theoreical Physics. 1970;**11**(4):220-222

[45] Ambartzumian RV, Letokhov VS, Ryabov EA, Chekalin NV. Isotopic selective chemical reaction of BCl_3 molecules in a strong infrared laser field. Journal of Experimental and Theoretical Physics Letters. 1974;**20**: 273-274

[46] Scruby RE, Lacher JR, Park JD. The infrared spectrum of boron trichloride. Journal of Chemical Physics. 1951;**19**(3): 386-387

[47] Wolfe DF, Humphrey GI. Force constants for pure and mixed halides of boron. Journal of Molecular Structure. 1969;**3**:293-303

[48] Maier WB II, Holland RF. High resolution infrared absorption spectra of BCl_3 and BCl_2F dissolved in solid argon and krypron. Journal of Chemical Physics. 1980;**72**(12):6661-6677

[49] Li Y-J et al. Exploration of laser-induced isotope separation of BCl_3. Chemical Physics Letters. 2021;**781**: 138948

[50] Bagratashvili VN, Letokhov VS, Makarov AA, Ryabov EA. Multiple

Photon Infrared Laser Photophysics and Photochemistry. London, Paris, New York: Harwood Academic Publishers; 1985

[51] Pirumov UG, Roslyakov GS. Gas Dynamics of Nozzles. Moscow: Main Office of Physics and Mathematical Literature Publishing Company; 1990

[52] Oerlikon Leybold Vacuum. Turbovac and MAG turbomolecular vacuum pumps. 2010

[53] Magnetically levitated turbo molecular pump instruction manual. 2006

Chapter 6

Effect of Capping Agents on the Nanoscale Metal Borate Synthesis

Fatma Tugce Senberber Dumanli

Abstract

Boron compounds are beneficial additives for industrial applications due to their superior physical, chemical, mechanical, and thermal features. The common use of boron compounds can be listed as ceramic, glass, glazes, metallurgy, lubricating agents, non-linear optical devices, and nuclear processes. Metal borates can be classified in accordance with the metal atom in the structure. According to the metal borate type, each compound exhibits different properties and is preferred for various applications. The other significant factor of a material that makes it preferable for the industry is its morphological characteristics. With the developing technology and novel synthesis procedures, metal borates can be fabricated at different morphologies. The characteristics of the metal borates can be improved by the modification of their surfaces. Capping agents are additive materials that are used to control particle growth and/or modify the morphological features of compounds. There is a recent increase in the number of studies based on metal borates prepared by using capping agents. In this chapter, the theoretical background on metal borates, synthesis procedures of metal borates, classification of the capping agent, the effect of capping agent on particle growth and examples of capping agent use on metal borates preparation were explained. Also, the characteristics of the same metal borates at different morphological features were compared.

Keywords: borate, capping agent, nanoparticle, morphology, synthesis, particle growth

1. Introduction

Boron is a rare portion element of earth's crust, and it is generally found in nature as the complexes of oxygen (O), hydrogen (H) other metal atoms. More than 150 boron minerals have been identified until today [1, 2]. Since including similar composition of most of the natural boron reserves (B, O, H, and metal atoms), "boron minerals" and "metal borates" might be considered synonymous words.

The majority of the boron reserves are found in Turkey (72.8%), Russia (7.6%), and South America (6.1%). Also, the smaller reserves could be seen in China, Kazakhstan, Argentina, Italy, Mexico, and Germany. The most abundant examples of the boron minerals can be listed as Borax ($Na_2B_4O_7{\cdot}10H_2O$), Tincalconite ($Na_2B_4O_7{\cdot}5H_2O$), Colemanite ($Ca_2B_6O_{11}{\cdot}5H_2O$), and Ulexite ($NaCaB_5O_6(OH)_6{\cdot}5H_2O$). In the reserves of boron minerals, particles commonly form in microcrystalline [3, 4].

The unique properties of borates could be explained by the high constant elasticity, heat resistance, corrosion resistance, luminescence, and low softening and melting temperatures [5, 6]. Also, recent studies exhibit the biocompatible properties of boron minerals [7, 8]. In the uses of boron compounds, the chemical nature and structure of borates provide multifunctionality [9]. Traditional uses of boron minerals can be seen in ceramics and glazes, detergents, agriculture, metallurgy, and fire retarding materials. With the effect of developed technologies, these compounds can also be utilized in energy storage systems, laser systems, optoelectronics, band gap engineering, tissue engineering, wound healing, bone regeneration, bone formation, antibacterial compositions, adsorption of pollutants from wastewater, and design of biochemical sensors [3–11].

Considering the properties that boron provides to the materials in which it is added and/or doped, its strategic importance emerges. For this reason, studies on the modification of metal borates, synthesis of novel metal borates, and related developments continue around the world. To increase their characteristics, specific types of metal borate compounds such as lanthanum borates or other rare earth-doped borates could be synthesized, or capping agents could be employed to modify their properties.

In the synthesis of specific metal borates, the reserves can be transformed to compounds at higher with high added value by the reaction metal sources with the boron mineral or boric acid (H_3BO_3) [12–14]. Lithium borates, zinc borates, and aluminum borates can be given as examples.

Capping agents could especially be used in nanoparticle synthesis. In the use of the capping agents, the particle sizes were mostly decreased, and the morphology was homogenized without changing experimental procedures. Also, produced particles become stabilized in the solution and agglomeration can be eliminated [15, 16]. Because of the high surface area to volume ratio of the smaller particle sizes, especially on the nanoscale, the compounds exhibited novel and remarkable features different from their bulk and molecular counterparts [17–19].

The characteristics of the prepared sample are associated with the designed experimental setup. For the effective use of capping agents, their role of them in the particle growth mechanism should be clarified. A detailed understanding of the connection between characteristics and synthesis procedures of the novel borates obtained will increase the correct form of metal borate use in industrial applications.

2. Capping agents

Capping agents are important additive materials to modify particle shape and size. The uses of the capping agents and modifiers have gradually increased both liquid-state and solid-state conditions. Several types of surfactants, polymers, extracts, ligands, cyclodextrins, dendrimers, and polysaccharides could be utilized with this aim. These organic compounds exhibit the ability to modify the metal surface, provide sufficient dispersion, and prevent the agglomeration of nanoparticle [20, 21].

2.1 Mechanism

The relationship between the capping agent and the metal surface is related to the electrical forces between the steric features of the surface, interfaces, composition, and chemical features of the ligand [20]. For a typically liquid-state synthesis procedure, the stages of particle formation include nucleation and growth. The modifying additives of capping agents are effective in both stages. Based on its characteristics, controlling the shape and size of the synthesized particles could be possible. With the addition of the capping agents, a draft could be prepared for the formation of the nucleus. The assumed draft also would be affected by the length of the molecules of the capping compound.

The growth mechanism of the particle can be explained with the steps of diffusion from cluster to particle surface and the bonding and/or reaction between ion and solid particles. Controlling the reaction and/or diffusion rate is also possible with the use of the reaction parameters [22].

Amphiphilic molecules of the capping agents include an apolar hydrocarbon group and a polar head group. The functionality of the capping agent is related to these apolar-polar groups. The apolar group reacts with the liquid medium, whereas the polar group bonded to the metal ion to contribute to the nanostructure [23]. The role of the capping agents on nanoparticle formation is schematized in **Figure 1**.

The capping molecule acts as a barrier for the transferring of ions on the produced particle surface. However, the partial transfer of the ions could be possible in the solution medium [22]. The selectivity of the capping agent is effective on the particle size of the prepared sample. The increasing selectivity will lead to lower mass transfer and limited particle growth. Two essential factors that affect the mass transfer between solution and particle are (i) the adsorption/desorption in the bulk and surfactant system; and (ii) the connection between the surfactant and the surface of the solid particle.

2.2 Classification

The functional groups of the capping agents are effective in the formation of solid particle–ligand interface, and ligand-solution interface. These groups are commonly

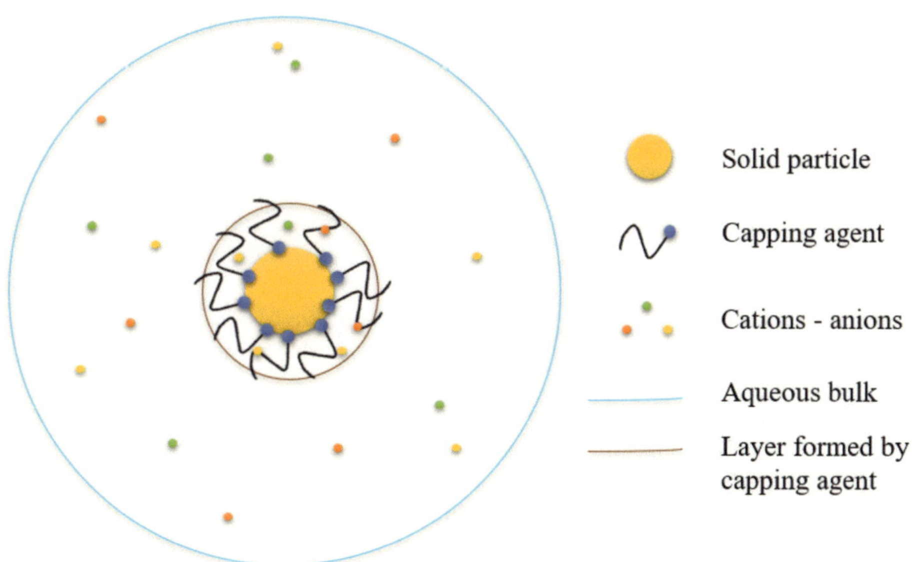

Figure 1.
The role of the capping agents on nanoparticle formation.

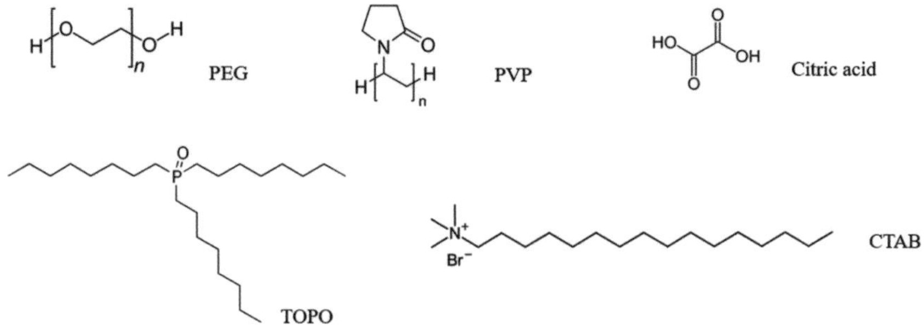

Figure 2.
Common examples of the capping agent with the different donor atoms.

found in polyatomic structures such as carboxylate, amino, and other coordinating groups including heteroatoms. According to the obtained interface between solid particle–ligand and ligand- solution, the fabricated sample could exhibit different properties such as hydrophilicity [24]. Anions or neutral molecules bound to the organic ligand centre are called donor atoms. The most common examples of the capping agent with the different donor atoms are presented in **Figure 2**.

The classification of the capping agents is based on the donor atom of capping agents [23, 25]:

- N-terminated capping agents are the organic ligands that include the donor atom of N such as cetyl trimethyl ammonium bromide (CTAB), octadecyl amine, (ODA), oleyamine, (oAm), hexadecyl amine, (HDA), polyvinyl pyrrolidone (PVP), and pyridine. These organic compounds are favorable to use in the preparation of metal oxide nanoparticles for being liquid state at room conditions, low cost, and stable in colloidal solutions.

- O-terminated capping agents are the ligands that include the donor atom of O such as polyethylene glycol, linolenic acid, and oleic acid. These compounds could react with the double bond or are absorbed by the carboxylate group of the nanoparticle.

- P-terminated capping agents are the organic structures that include the donor atom of P such as tri-n-octyl phosphine (TOP) and tri-n-octyl phosphene oxide (TOPO). Especially, they are preferable in the hot injection preparation methods with the starting materials of organometallic compounds.

- S-terminated capping agents are the organic ligands that include the donor atom of S such as 1-thioglycerol. Mostly, they exhibit hydrophilic features.

- Green capping agents are the ligands such as citric acid, enzymes, polyphenols, and biodegradable polymers. There is a remarkable increase in the use of green capping agents especially in nanoparticle synthesis with the development of green technologies, and the increasing importance of sustainability. The green ligands can be easily obtained from the extraction of aromatic plants, fruits, roots, leaves, and their peels. In the preparation of green ligands, the most common solvent used in the extraction process is water. The biochemical composition amount and quality may be affected by the operation parameters of the extraction process. However, there is no research was seen on the use of green capping agents in the modification of metal borates. This may be due to the effect of the capping agent on the particle growth mechanism that has not been adequately studied for the metal borate modification.

3. Metal borate compounds

3.1 Structure of the metal borates

Metal borates are well-known materials due to their superior properties such as their resistance to physical, chemical, and thermal conditions, the activities of magnetic and electrical, and antibacterial behavior [12]. Metal atom links to the borate groups at the positions of tetragonal and trigonal, and the compounds exhibit symmetrical and asymmetrical stretching. These symmetric and asymmetric stretchings are typical for a molecule and could be examined by using the Fourier-transform infrared spectroscopy (FT-IR) and/or Raman spectroscopy. The characteristic band values of borates groups for metal borates for the FT-IR and Raman spectrums are summarized in **Table 1**.

Some properties of the metal borate could be intrinsic to the metal atom bonded in structure, the designed experimental procedure, and the crystallinity of the sample. This would lead to specific uses of the metal borates according to their types. The synthesis of modified boron minerals can be beneficial for the special uses of metal borates. The common uses of metal borates according to the metal atom bonded to the borate structure can be seen in **Table 2**.

The modification of the metal borates would also lead to novel applications. Although the traditional uses of magnesium borates are thermoluminescence and neutron shielding, the studies on their novel uses showed that magnesium borates modified by capping agents exhibited the hierarchical porous microspheres in morphology and could be an alternative for the Congo Red adsorption from wastewater [37 – 39].

3.2 Synthesis procedures

Metal borates could be produced by using both liquid-state (hydrothermal) and solid-state (thermal) methods. The synthesis procedure of metal borate could also be adapted to novel technologies such as ultrasound and microwave methods [40, 41]. Although these technologies provide a minor or major decrease in particle size distribution, there is still a requirement for the homogenization in surface morphology of the obtained samples.

Stretching type		FT-IR (cm^{-1})	Raman (cm^{-1})
$B_{(3)}$-O	Asymmetrical	1600–1400	1500–1300
	Symmetrical	1400–1200	1300–1200
$B_{(4)}$-O	Asymmetrical	1200–950	1200–1050
	Symmetrical	950–700	1050–750
Polyanions of $(B_6O_7(OH)_6)^{-2}$ and $(B_3O_3(OH)_4)^{-2}$		700–600	750–620

Table 1.
The characteristic band values of borates groups for metal borates for the FT-IR and Raman spectrums [26–28].

Type	Chemical name	Chemical formula	Uses	Ref.
Aluminum borates	Aluminum borate	$Al_4B_2O_9$	Ceramic filters	[29]
Barium borates	Barium borate	BaB_2O_4	Non-linear optical devices	[13]
Calcium borates	Colemanite	$Ca_2B_6O_{11}{\bullet}5(H_2O)$	Ceramics and glazes	[12]
	Inyoite	$Ca_2B_6O_{11}{\bullet}13(H_2O)$		
	Pandermite	$Ca_4B_{10}O_{19}{\bullet}7(H_2O)$		
Copper borates	Santarosaite	CuB_2O_4	Electrical and magnetic applications	[30]
Iron borates	Vonsenite	$Fe^{2+}{}_2Fe^{3+}BO_5$	Magnetic and optical applications	[31]
Lanthanum borates	Lanthanum borate	$LaBO_3$	Non-linear optical devices	[32]
Lithium borates	Diomignite	$Li_2B_4O_7$	Ion batteries, opto-electronic materials	[33]
Magnesium borates	Inderite	$MgB_3O_3(OH)_5{\bullet}5(H_2O)$	Thermoluminescence	[34]
	Admontite	$Mg(B_6O_{10}){\bullet}7(H_2O)$		
	Mcallisterite	$Mg_2B_{12}O_{14}(OH)_{12}{\bullet}9(H_2O)$		
Potassium borates	Santite	$KB_5O_6(OH)_4{\bullet}2(H_2O)$	Non-linear optical devices	[35]
Sodium borates	Borax	$Na_2(B_4O_5)(OH)_4{\cdot}8(H_2O)$	Glass and metallurgy	[2]
	Tincalconite	$Na_2[B_4O_5(OH)_4]{\cdot}3H_2O$		
Zinc borates	Zinc borate	$Zn_3B_6O_{12}{\cdot}3.5H_2O$	Fire retardant materials	[36]

Table 2.
Examples, chemical formulas, and uses of some metal borates [2, 12, 13, 29–36].

Liquid-state synthesis of metal borates involves the stages of (i) dissolution of metal and boron sources in liquid mediums, (ii) mixing the prepared solutions at the suitable ratio, reaction temperature, and time, and (iii) filtration and drying if it is necessary. Water is selected as the solution medium in liquid-state conditions; however, other types of fuels could also be used in metal borate preparation. The liquid-state synthesis procedure could also be entitled as hydrothermal, combustion, and co-precipitation methods [42–44]. The capping agent is usually added to the mixture of the dissolved sources. The main point of the liquid-state synthesis procedure is that the selected capping agent should be soluble in the solution for better interaction between the core particle and the capping agent molecules.

Solid-state synthesis of metal borates includes the stages of (i) mixing of metal and boron sources at the suitable mole ratios at powder state, (ii) calcination of prepared mixtures at the suitable reaction temperature and time, (iii) grinding, if it is necessary. The described method can be entitled as the solid-state, calcination or thermal method [45–47]. The capping agent should be added to the powder mixture. The main point of the solid-state synthesis procedure is to obtain a homogeneous mixture of powder at the initial state.

To improve the characteristics of the metal borates, some experimental procedures could also be defined as the combination of both methods. Most of the combinations include the preparations of metal and boron complexes in hydrothermal conditions and the calcination of the prepared complexes [44].

3.2.1 Effects of capping agents on the produced metal borates

In the heterogeneous form of processes of adsorption, desorption, surface reaction, and adsorbate lateral diffusion, each process includes bond-breaking and bond-making facts, which are related to the electronic properties of the reactants and surfactant material [20]. The growth-limiting role of the capping agent on particles is also based on the relationship between the particle and the capping agent. The efficiency of the capping agent is mainly based on the suitable matching of material and reaction conditions and the interaction of the capping agent with the core particle. Commonly, the proper amount of capping agent in the reaction medium is assumed as 1% or lower.

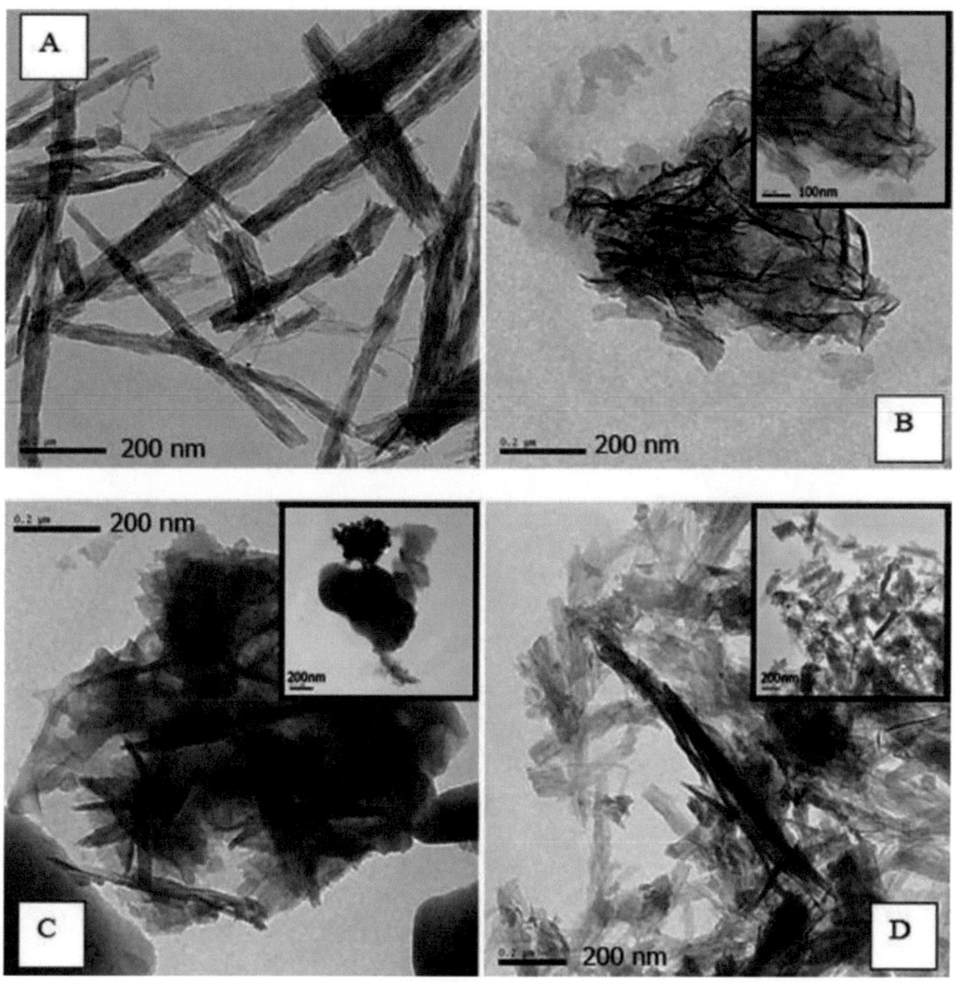

Figure 3.
The modification of hydrothermally synthesized zinc borates by using (A) without capping agent, (B) CTAB added, (C) triton 114 added, and (D) oleic acid added [36].

The advantages of the capping agent in metal borate synthesis can be listed as the homogenization of the sample surface, modification of particle shape, and obtaining smaller particle size. By using the cationic surface-active agent of CTAB, nonionic surface-active agent of Triton-114, and anionic surface-active agent of oleic acid, the modification of the surfaces of the hydrothermally synthesized zinc borates was presented in **Figure 3**. As could be seen in the SEM analysis results, in **Figure 3**, the capping agent uses of the CTAB and oleic acid decreased the length of the prepared samples whereas the use of T as a capping agent reshaped the particle appearance. Ipek (2020) also indicated the improvement of the fire-retardant properties of zinc borate by decreasing the particle sizes with the use of capping agents [36]. To overcome the problem of incompatibility of zinc borate in polymer matrixes, Li et al. (2010) produced hydrophobic zinc borate by the modification with oleic acid [47].

Erfani et al. (2012) synthesized ultra fine particles of calcium tetraborate by using the capping agent of PVP to control the particle size and to reduce the agglomeration in a co-precipitation method [17]. Khalilzadeh et al. (2016) experimented that increasing PVP addition up to a certain value in the production of lithium tetraborate narrowed the particle size range, reduced the average particle diameter, decreased agglomeration, and increased the band gap value [33]. Xing et al. (2019) studied the modifying effect of folic acid on the yttrium borate to strengthen its photoluminescence emission intensity features in hydrothermal conditions [48]. Liu et al. (2010) synthesized the nanoscale europium-doped barium borate (Ba–B–O: Eu^{3+}) with the addition of oleic acid as a capping agent and enhanced the emission intensities of products by adjusting the correct ratio of the modifying agent [13].

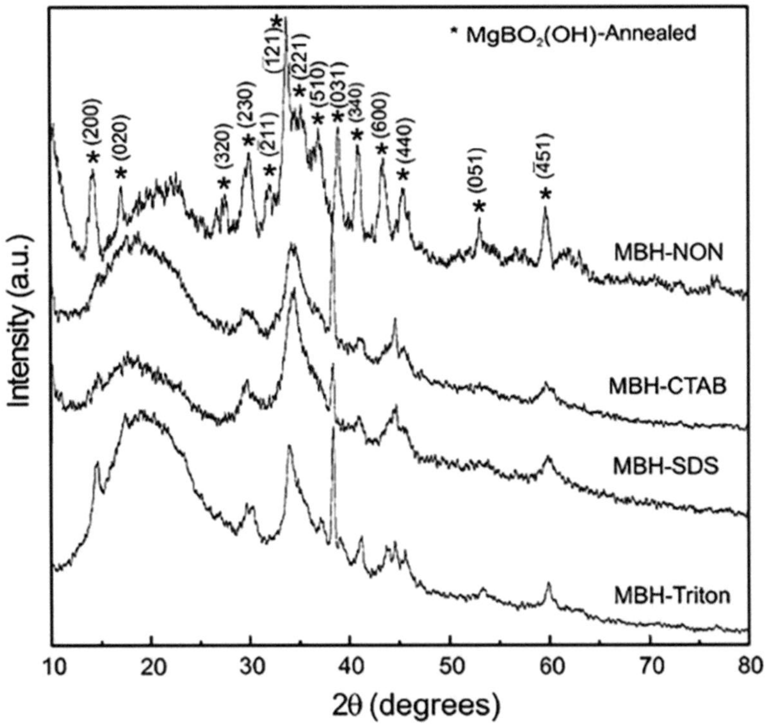

Figure 4.
XRD results of the synthesized magnesium borate hydroxides with different capping agents [49].

The only drawback of the capping agent usage in metal borate synthesis can be explained by the decrease in crystallinity. A few studies indicated that the use of a capping agent could adversely affect the prepared sample and a decrease in the crystallinity of the product could be seen in some cases. In the fabrication of magnesium borate hydroxide powders, Kumari et al. studied the effects of different capping agents CTAB, SDS, and Triton on the products. The results indicated the well-modified morphology of the sample at lower crystallinity [49]. The XRD results of Kumari et al. were presented in **Figure 4**. As it was given in **Figure 4**, the adverse effect of capping agents on the Triton, SDS and CTAB added magnesium borate hydrates can be seen with the decreasing peaks of the (h k l) values of (0 2 0), (3 2 0), (2 1 1), (2 2 1), (5 1 0), (3 4 0) and (0 5 1). The decrease in crystallinity could be explained by the limited growth of the core particles due to the capping agent effect.

4. Conclusion

With the help of the nanoscale metal borate productions with the usage of the capping agents, samples with high added value have been obtained compared to the microcrystalline structures. In the use of a capping agent, the type of the capping agent, the amount, being soluble in the selected solvent and the interaction between the capping agent and particle are the key points. For the correct determination of the capping agent, the particle growth mechanism, and effects of the capping agent on it were detailed studied. Also, the advantages and disadvantages of the capping agent usage in metal borate preparation were discussed. The novel uses and changes in characteristics of modified nanoscale metal borates were presented.

It is expected that the significance of metal borates in industrial applications would be expanded with the increase of advanced technologies in their synthesis. In this case, the development of modified synthesis techniques with a novel experimental setup is suggested.

Nomenclature

CTAB	Cetyl trimethyl ammonium bromide
FT-IR	Fourier-transform infrared spectroscopy
HDA	Hexadecyl amine
oAm	Oleyamine
ODA	Octadecyl amine
PEG	Polyethylene glycol
PVA	Polyvinyl alcohol
PVP	Polyvinyl pyrrolidone
SDS	Sodium dodecyl sulphate
SEM	Scanning Electron Microscope
T	Triton
TOP	tri-n-octyl phosphine
TOPO	tri-n-octyl phosphene oxide
XRD	X-Ray Diffraction

Author details

Fatma Tugce Senberber Dumanli
Nisantasi University, Istanbul, Turkey

*Address all correspondence to: fatma.senberber@nisantasi.edu.tr

References

[1] Adair R. Understanding the Elements of Periodic Table: Boron. 1st ed. New York: The Rosen Publishing Group, Inc.; 2007. p. 7

[2] Powoe SPB, Kromah V, Jafari M, Chelgani SC. A review on the beneficiation methods of borate minerals. Minerals. 2021;**11**(318):1-18. DOI: 10.3390/min11030318

[3] State Planning Organization of Tukey Ninth Development Plan (2007-2013), Chemical Industry Specialization Commission, Boron Working Group Report. Available from: https://www.sbb.gov.tr/wp-content/uploads/2022/07/Dokuzuncu_Kalkinma_Plani-2007-2013.pdf [Accessed: April 23, 2023]

[4] The Union of Chambers of Turkish Engineers and Architects (TMMOB) – Boron Report. Available from: https://www.tmmob.org.tr/sites/default/files/bor_0.pdf [Accessed: April 20, 2023]

[5] Bengisu M. Borate glasses for scientific and industrial applications: A review. Journal of Materials Science. 2016;**51**:2199-2242. DOI: 10.1007/s10853-015-9537-4

[6] Keszler D. Synthesis, crystal chemistry, and optical properties of metal borates. Current Opinion in Solid State and Materials Science. 1999;**4**:155-162. DOI: 10.1016/S1359-0286(99)00011-X

[7] Ege D, Zheng K, Boccaccini AR. Borate bioactive glasses (BBG): Bone regeneration, wound healing applications, and future directions. ACS Applied Biomaterials. 2002;**5**:3608-3622. DOI: 10.1021/acsabm.2c00384

[8] Hawkins EH, Teixidor CV, editors. Boron-Based Compounds: Potential and Emerging Applications in Medicine. 1st ed. Chichester: Wiley; 2018. p. 24

[9] Shah FU, Glavatskih S, Antzutkin ON. Boron in tribology: From borates to ionic liquids. Tribology Letters. 2013;**51**:281-301. DOI: 10.1007/s11249-013-0181-3

[10] Fauzillah NA, Abdullah I, Ivandini TA. Modification of boron-doped diamond electrode with gold nanoparticles synthesized by allyl Mercaptan as the capping agent for arsenic sensors. AIP Conf. Proc. 2020;**2242**:040031. DOI: 10.1063/5.0008495

[11] Yang J, Sun X. Borate particulate photocatalysts for photocatalytic applications: A review. International Journal of Hydrogen Energy. 2022;**47**:5608-5630. DOI: 10.1016/j.ijhydene.2022.05.305

[12] Helvaci C. Borate deposits: An overview and future forecast with regard to mineral deposits. Journal of Boron. 2017;**2**:59-70

[13] Liu N, Zhao D, Yu L, Zheng K, Qin W. Controlled synthesis and photoluminescence of europium doped barium borate nanorods, nanowires, and flower-like assemblies. Colloids and Surfaces A: Physicochemical and Engineering Aspects. 2010;**363**:124-129. DOI: 10.1016/j.colsurfa.2010.04.028

[14] Yalcin A, Gonen M. Lithium tetraborate production from the reaction of boric acid and lithium carbonate using carbon dioxide. Sigma Journal of engineering and Natural Sciences. 2020;**38**:1121-1132

[15] Javed R, Sajjad A, Naz S, Sajjad H, Ao Q. Significance of capping agents

of colloidal nanoparticles from the perspective of drug and gene delivery, bioimaging, and biosensing: An insight. International Journal of molecular Sciences. 2022;**23**:10521. DOI: 10.3390/ijms231810521

[16] Rajpoot Y, Sharma V, Basak S, Ali W. Calcium borate particles: Synthesis and application on the cotton fabric as an emerging fire retardant. Journal of Natural Fibers. 2022;**19**:5663-5675. DOI: 10.1080/15440478.2021.1889431

[17] Erfani M, Saion E, Soltani N, Hashim M, SaffieyW AB, Navasery M. Facile synthesis of calcium borate nanoparticles and the annealing effect on their structure and size. International Journal of molecular Sciences. 2012;**13**:14434-14445. DOI: 10.3390/ijms131114434

[18] Haghiri ME, Soltani N, Saion E, Izanloo A, Bahmanrokh G, Askari M. Fabrication and radiation dose properties ofWell-dispersed calcium borate nanoparticles. Nanoscience & Nanotechnology-Asia. 2019;**9**:198-209. DOI: 10.2174/2210681207666171107142 45

[19] Rajendran K, Sen S. Effect of capping agent on antimicrobial activity of nanoparticles. Der Pharmacia Lettre. 2015;7:37-42

[20] Campisi S, Schiavoni M, Chan-Thaw CE, Villa A. Untangling the role of the capping agent in Nanocatalysis: Recent advances and perspectives. Catalysts. 2016;**6**:185. DOI: 10.3390/catal6120185

[21] Yang TH, Shi Y, Janssen A, Xia Y. Surface capping agents and their roles in shape controlled synthesis of colloidal metal nanocrystals. Angewandte Chemie International Edition. 2019;**59**:15378-15401. DOI: 10.1002/anie.201911135

[22] Phan CM, Nyungen HM. Role of capping agent in wet synthesis of nanoparticles. The Journal of Physical chemistry A. 2017;**121**:3213-3219. DOI: 10.1021/acs.jpca.7b02186

[23] Gulati S, Sachdeva M, Bhasin KK. Capping agents in nanoparticle synthesis: Surfactant and solvent system. AIP Conference Proceedings. 2018;**1953**:030214. DOI: 10.1063/1.5032549

[24] Arulmozhi K, Mythili N. Studies on the chemical synthesis and characterization of lead oxide nanoparticles with different organic capping agents. AIP Advances. 2013;**3**:122122. DOI: 10.1063/1.4858419

[25] Javed R, Zia M, Naz S, Aisida SO, Ain N, Ao Q. Role of capping agents in the application of nanoparticles in biomedicine and environmental remediation: Recent trends and future prospects. Journal of Nanobiotechnology. 2020;**18**:172. DOI: 10.1186/s12951-020-00704-4

[26] Frost RL, Lopez A, Xi Y, Lima RMF, Scolz R, Granja A. The molecular structure of the borate mineral inderite $Mg(H_4B_3O_7)(OH)\cdot 5H_2O$ – A vibrational spectroscopic study. Spectrochimica Acta Part A: Molecular and Biomolecular Spectroscopy. 2013;**116**:160-164. DOI: 10.1016/j.saa.2013.06.108

[27] Gautam C, Yadav AK, Singh AK. A review on infrared spectroscopy of borate glasses with effects of different additives. International Scholarly Research Notices. 2012;**2012**:428497. DOI: 10.5402/2012/428497

[28] Liu Y, Deng JX, Pan CY. Synthesis and structural manipulation of a Polyanionic Zincoborate $[Zn(H_2O)_6](NO_3)_2\cdot [Zn\{B_3O_3(OH)_4\}2(H2O)_4]\cdot 2H_2O$ by the molten salt method and its

potential application as a flame retardant. Journal of Cluster Science. 2021;**32**:469-477. DOI: 10.1007/s10876-020-01806-7

[29] Dai H, Gong J, Kim H, Lee D. A novel method for preparing ultra-fine alumina-borate oxide fibres via an electrospinning technique. Nanotechnology. 2002;**13**:674-677. DOI: 10.1088/0957-4484/13/5/327

[30] Kipcak AS, Senberber FT, Yuksel SA, Derun EM, Piskin S. Synthesis, characterisation, electrical and optical properties of copper borate compounds. Materials Research Bulletin. 2015;**70**:442-448. DOI: 10.1016/j.materresbull.2015.05.003

[31] Cam I, Timucin M. Chemical synthesis of iron borate, Fe_3BO_6. In: 2nd International Boron Symposium, 23-25 September 2004 Eskişehir Turkey. 2004

[32] Jia Z, Xia Y. Hydrothermal synthesis, characterization, and Tribolo gical behavior of oleic acid-capped lanthanum borate with different morphologies. Tribology Letters. 2011;**41**:425-434. DOI: 10.1007/s11249-010-9728-8

[33] Khalilzadeh N, Saion EB, Mirabolghasemi H, Shaari AHB, Hashim MB, Ahmad MBH, et al. Single step thermal treatment synthesis and characterization of lithium tetraborate nanophosphor. Journal of Materials Research and Technology. 2016;**5**:37-44. DOI: 10.1016/j.jmrt.2015.05.005

[34] Senberber Dumanli FT. Magnesium borates: The relationship between the characteristics, properties, and novel technologies. In: Current Trends in Magnesium (Mg) Research. IntechOpen; 2022. pp. 133-150. DOI: 10.5772/intechopen.104487

[35] Asensio MO, Yildirim M, Senberber FT, Kipcak AS, Derun EM. Thermal dehydration kinetics and characterization of synthesized potassium borates. Research on Chemical Intermediates. 2016;**42**:4859-4878. DOI: 10.1007/s11164-015-2326-5

[36] Ipek Y. Effect of surfactant types on particle size and morphology of flame-retardantzinc borate powder. Turkish Journal of Chemistry. 2020;**44**:214-223. DOI: 1 0.3906/kim-1906-49

[37] Bahl S, Pandey A, Lochab SP, Aleynikov VE, Molokanov AG, Kumar P. Synthesis and thermoluminescence characteristics of gamma and proton irradiated nanocrystalline MgB4O7: Dy, Na. Journal of Luminescence. 2013;**134**:691-698. DOI: 10.1016/j.jlumin.2012.07.008

[38] Guo RF, Ma YQ, Liu ZH. Three hierarchical porous magnesium borate microspheres: A serial preparation strategy, growth mechanism and excellent adsorption behavior for Congo red. RSC Advances. 2019;**9**:20009-20018. DOI: 10.1039/C9RA03654G

[39] Pellicioni M, Prokic M, Esposito A, Nuccetelli CP. Energy response of graphite-mixed magnesium borate TLDs to low energy X-rays. Applied Radiation and Isotopes. 1991;**42**(11):1037-1038

[40] Ila S, Kipcak AS, Derun EM. Ultra-fast and effective ultrasonic synthesis of potassium borate: Santite. Main Group Metal Chemistry. 2022;**45**:19-25. DOI: 10.1515/mgmc-2022-0004

[41] Ila S, Kipcak AS, Derun EM. A new rapid synthesis of potassium borates by microwave irradiation. Main Group Chemistry. 2020;**19**:265-272. DOI: 10.3233/mgc-200956

[42] Derun EM, Senberber FT. Characterization and thermal dehydration kinetics of highly crystalline Mcallisterite, synthesized

at low temperatures. The Scientific World Journal. 2014;**2014**:985185. DOI: 10.1155/2014/985185

[43] Fu L, Zhao X, Meng Z, Zhang J, Chu G. Characterization and luminescent properties of lanthanum borate prepared by solution combustion method. Materials Science and Engineering: B. 2023;**291**:116362. DOI: 10.1016/j.mseb.2023.116362

[44] Argun S, Senberber Dumanli FT, Kipcak AS, Tugrul N, Moroydor DE. Co-precipitation synthesis of lanthanum borates with different morphologies. Glass Physics and Chemistry. 2023;**9**:92-97. DOI: 10.1134/S108765962260020X

[45] Verma S, Verma K, Kumar D, Chaudhary B, Som S, Sharma V, et al. Recent advances in rare earth doped alkali-alkaline earth borates for solid state lighting applications. Physica B: Condensed Matter. 2018;**535**:106-113. DOI: 10.1016/j.physb.2017.06.073

[46] Kipcak AS, Acarali N, Senberber FT, Yildirim M, Turkmen S, Aydin Yuksel S, et al. Synthesis of dehydrated zinc borates using the solid-state method and characterization and investigation of the physical properties. Main Group Chemistry. 2016;**15**:301-313. DOI: 10.3233/MGC-160210

[47] Li S, Long B, Wang Z, Tian Y, Zheng Y, Zhang Q. Synthesis of hydrophobic zinc borate nano flakes and its effect on flame retardant properties of polyethylene. Journal of Solid State Chemistry. 2010;**183**:957-962. DOI: 10.1016/j.jssc.2010.02.017

[48] Xing X, Li S, Song Y, Ge Y, Zhang X, Jiang W, et al. Novel hierarchical YBO_3: Eu^{3+} nanocrystals synthesized by folic acid assisted hydrothermal process. MRS Advances. 2019;**5**:1661-1670. DOI: 10.1557/adv.2019.455

[49] Kumari L, Li WZ, Kulkarni S, Wu KH, Chen W, Wang C, et al. Effect of surfactants on the structure and morphology of magnesium borate hydroxide nanowhiskers synthesized by hydrothermal route. Nanoscale Research Letters. 2010;**5**:149-157. DOI: 10.1007/s11671-009-9457-9

Chapter 7

Numerical Simulation of the Mechanical Behaviour of Boron Nitride Nanosheets and Nanotubes

Nataliya A. Sakharova, Jorge M. Antunes, André F.G. Pereira, Bruno M. Chaparro and José V. Fernandes

Abstract

Among the compounds formed by an element of the 13th group and nitrogen, boron nitride, also called white graphene, stands out for its high strength and thermal conductivity, transparency to visible light, antimicrobial properties, high resistance to oxidation, and biocompatibility. One-dimensional and two-dimensional boron nitride nanostructures, i.e. nanotubes and nanosheets, respectively, are expected to present innovative advanced characteristics not equal to those of bulk boron nitride, bringing new perspectives to numerous applications in nanoscale electronics and biomedicine. For the correct design of systems and devices consisting of boron nitride nanosheets and nanotubes, understanding the mechanical behaviour of these nanostructures is extremely important. Firstly, because the robustness and functioning of nanosystems and nanodevices based on boron nitride nanostructures are determined by the mechanical behaviour of their constituents and also because deformation can influence the optical, electric, and thermoelectric properties of boron nitride nanotubes and nanosheets. In this context, the current chapter is dedicated to the numerical evaluation of the elastic properties of boron nitride nanosheets and nanotubes, using the nanoscale continuum modelling (also called molecular structural mechanics) approach. With this aim, a three-dimensional finite element model was used to evaluate their elastic moduli.

Keywords: boron nitride, nanosheets, nanotubes, elastic moduli, modelling, numerical simulation

1. Introduction

Hexagonal boron nitride (h-BN) is one of the compounds of the 13th–15th groups, consisting of alternating boron (B) and nitrogen (N) atoms, linked by strong covalent bonds, forming a honeycomb lattice structure similar to that of graphene. For this reason, h-BN is often called "white graphene" [1, 2]. Hexagonal BN is a promising ultra-wide bandgap semiconductor with notable physical properties, chemical and thermal stability, and compatibility with graphene [1–5]. Two-dimensional (2D) h-BN layers have been broadly integrated into van der Waals heterostructures, with the aim to develop novel architectures for 2D electronic devices [5, 6]. The combination of 2D boron nitride layers with those of graphene enables the emergence of new applications and improves the performance of conventional devices. Dean et al. [3] demonstrated that the use of h-BN substrates instead of standard silica brings advantages to high-quality graphene electronics, allowing to reduce roughness and chemical reactivity of graphene devices. Among the potential applications envisaged in recent times for 2D BN, resonant nanoelectromechanical systems (NEMS) [7], membranes for the separation of toxic gases [8] and water desalination [9], quantum emitters [10], anodes for magnesium-ion batteries [11]. Considering the diversity of potential applications, the development of methods for synthesised BN nanosheets has received special research attention (see, for example [1, 2, 12–15]). Hexagonal BN nanosheets with few atomic layers were synthesised using the thermal chemical vapour deposition (CVD) method on copper substrate [1, 2], polycrystalline nickel and cobalt substrates [2], and amorphous silica and quartz substrates [13]. The method proposed by Song et al. [1] allowed producing h-BN nanosheets at a large scale. Obtaining 2D BN nanosheets by micromechanical cleavage, a technique that consists of peeling off h-BN layers from bulk crystals, was also reported [3, 12, 15]. Moreover, a low-cost, template-free method for chemical synthesis of single-layer h-BN areas, using different acids as precursors, was recently proposed [14].

Single-walled boron nitride nanotubes (NTs), 1D structures, which can be understood as rolled-up BN nanosheets, are in the research focus due to their perspectives in highly demanded applications such as membranes for gas separation [16] and water filtration [17], osmotic power generators [18], and nanosensors [19]. The structural similarity of boron nitride nanotubes (BNNTs) and carbon nanotubes (CNTs) motivated research in order to replace CNTs with BNNTs in existing applications. Among these, there are the reinforcement of nanocomposites and the improvement of advanced functional materials. But innovative applications have also been sought, where the combination of boron nitride and carbon nanotubes in hybrid nanostructures permits taking advantage of both constituents and achieving better properties and performance of novel devices. After Chowdhury and Adhikari [20] first synthesised boron nitride nanotubes using arc discharge, in 1995, BNNTs were successfully produced by ball-milling [21–23], CVD [24–27], laser ablation [28–32], and thermal plasma jet [33–36] methods. Regarding the intended applications for 1D and 2D boron nitride nanostructures, their mechanical properties play a significant role in reinforced composites, membranes, van der Waals structures, as well as in the functioning and strength of nanosystems and nanodevices. On the other hand, deformation can modify the electronic, thermoelectric, optical, and chemical properties of boron nitride 1D and 2D nanostructures [37, 38].

The studies devoted at evaluating the mechanical properties of boron nitride nanosheets (BNNSs) and BNNTs are mostly carried out theoretically because experimental techniques at the nanoscale require many resources that are expensive. There are three classes of theoretical approaches (analytical and numerical) used to model and characterise the mechanical behaviour of BNNSs and BNNTs, namely the atomistic, the continuum mechanics (CM), and the nanoscale continuum modelling (NCM) approaches, the latter also called molecular structural mechanics (MSM). With regard to the atomistic approach, there are studies where the elastic properties of BNNSs and BNNTs were evaluated using *ab initio* Density Functional Calculations (DFT) and molecular dynamics (MD). The latter makes use of different potential functions to describe the interactions between B and N atoms in the nanostructures. Ahangari et al. [39] calculated Young's modulus of BNNSs by the *ab initio* DFT method, and Le [40] and Zhao and Xue [41] used MD simulations with Tersoff potentials for this purpose. Concerning the BN nanotubes, Santosh et al. [42] used MD simulation, with force–constant approach to describe the interaction between B and N atoms, in order to calculate the Young's and shear moduli of the BNNTs. Tao et al. [43] performed MD simulation with Tersoff–Brenner (TB) potential, combining it with the finite element (FE) method to evaluate Young's modulus of BNNTs. Kumar et al. [44] in their MD simulation study used a second-generation reactive empirical bond order (REBO) potential and evaluated the BNNTs Young's modulus. The use of the CM approach, which models the entire nanostructure as a continuum element, is not frequent in the literature. Oh [45] made use of the CM approach combined with the adjusted TB potential to evaluate Young's modulus of the BNNTs.

Unlike the CM approach, the NCM/MSM considers the bonds between B and N atoms as elastic elements (springs or beams), taking advantage of the connection between the molecular structure of the nanosheet or nanotube and the solid mechanics. Tapia et al. [46] used the NCM/MSM approach, in which the B-N bond is replaced by the beam element, to evaluate Young's and shear moduli of the BNNSs. The NCM/MSM approach has often been chosen as the most suitable for efficient and fast computational simulation of the mechanical response of BNNTs. Genoese et al. [47] evaluated the Young's and shear moduli of BNNTs, relating the "stick-and-spring" (NCM/MSM) and the Donnell (CM) model of continuum thin shell. Li and Chou [48] and Ansari et al. [49] replaced the B-N bond with the beam element for evaluating of BNNTs elastic properties under NCM/MSM approach, while Zakaria [50] used two-sectioned beam elements for this purpose. Giannopoulos et al. [51], instead of beam elements, modelled the B-N bond as spring-like elements to calculate the BNNTs Young's modulus. Yan et al. [52] studied longitudinal and torsional free vibrations of BNNTs in a framework of the NCM/MSM approach and derived analytical solutions for Young's and shear moduli.

Although most achievements in evaluating the elastic properties of boron nitride nanosheets (NSs) and nanotubes (NTs) are due to theoretical approaches, many experimental studies have been carried out for this purpose. Song et al. [1] measured Young's modulus of BNNSs with 2–5 layers, using the nanoindentation test carried out by atomic force microscopy (AFM). Arenal et al. [53] used *in situ* uniaxial compression tests performed by high-resolution transmission electron microscopy (HRTEM) and AFM to evaluate Young's modulus of single-walled boron nitride nanotubes (SWBNNTs). Chen et al. [54] evaluated the Young's modulus of multi-walled boron nitride nanotubes (MWBNNTs), with the help of *in situ* axial compression, using a transmission electron microscope (TEM) with a force transducer holder. Chopra and Zettl [55] and Suryavanshi et al. [56] measured Young's modulus of MWBNNTs using the thermal vibrational amplitude of a cantilevered nanotube and the electric-field-induced resonance method, respectively, in TEM. Zhou et al. [57] used a high-order resonance technique within HRTEM for this purpose. Golberg et al. [58] and Ghassemi et al. [59] evaluated Young's modulus of the MWBNNTs from *in situ* bending and cycling bending tests, respectively, carried out with the help of AFM set up within TEM. Tanur et al. [60], with a resource to a three-point bending technique in AFM, calculated Young's and shear moduli of MWBNNTs.

The aim of the present study is to evaluate elastic moduli of one-layer boron nitride nanosheets (BNNSs) and non-chiral single-walled boron nitride nanotubes (SWBNNTs) employing the NCM/MSM approach with beam elements to simulate B-N covalent bond. For this purpose, 3D FE model was used, which permits evaluating Young's and shear moduli of 1D and 2D boron nitride nanostructures. The Young's modulus results obtained for SWBNNTs are analysed, bearing in mind those for BNNSs. Moreover, the current study presents a reference to determine the elastic properties of BNNSs and SWBNNTs and the architecture of hybrid 1D and 2D structures based on carbon and boron nitride.

2. Atomic structure of BNNSs and SWBNNTs

The BNNS has a honeycomb atomic arrangement with planar geometry [61] as shown in **Figure 1**, where the chiral vector, $\boldsymbol{C_h}$, and the chiral angle, θ, are drawn. The unit vectors of the hexagonal BN lattice, $\boldsymbol{a_1}$ and $\boldsymbol{a_2}$, and the chiral indices, n and m (integers), are also represented in this figure. These parameters allow defining the chiral vector and chiral angle by the following expressions:

$$\boldsymbol{C_h} = n\boldsymbol{a_1} + m\boldsymbol{a_2}, \tag{1}$$

$$\theta = \sin^{-1}\frac{\sqrt{3}}{2}\frac{m}{\sqrt{n^2 + nm + m^2}}. \tag{2}$$

Rolling up the h-BN sheet into a cylinder, while varying the chiral angle, θ, from 0° to 30°, results in three types of SWBNNTs configurations, as exemplified in **Figure 1**. When θ = 0° ($m = 0$) and θ = 30° ($n = m$), the resulting tubular structures are called $(n, 0)$ zigzag and (n, n) armchair nanotubes, respectively, and form the symmetry group of non-chiral NTs (see, **Figure 2**). Intermediate NT configurations, for which 0° < θ < 30° ($n \neq m$), are related to the symmetry group of (n, m) chiral NTs.

The geometric structure of the SWBNNTs is described through the circumference, L_C, and the diameter, D_n, of the nanotube expressed as follows:

$$L_C = |\boldsymbol{C_h}| = a\sqrt{n^2 + nm + m^2}, \tag{3}$$

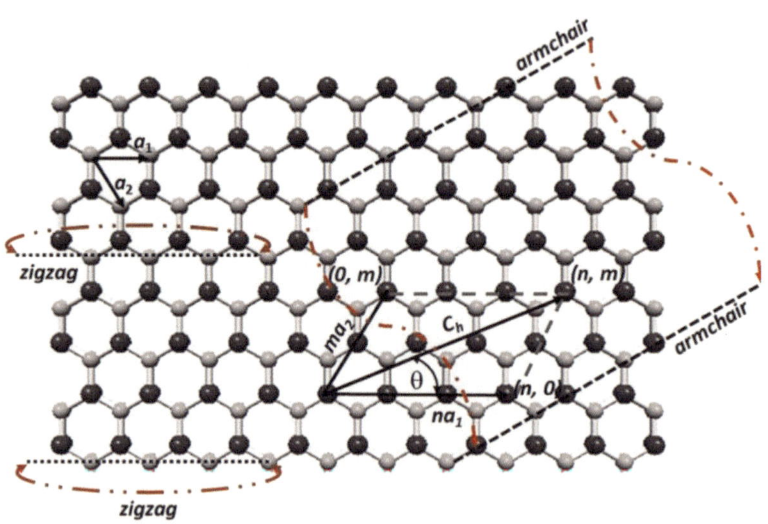

Figure 1.
Schematic illustration of the h-BN sheet with the chiral vector, $\boldsymbol{C_h}$, chiral angle, θ, and indication of rolling-up for zigzag and armchair BNNTs. B atoms are represented in black; N atoms are in grey.

Figure 2.
Zigzag and armchair configurations of SWBNNTs, built using the software Nanotube Modeler©. B atoms are represented in black; N atoms in grey.

Reference	[46]	[61]	[44]	[62]	[63]	[64]	[65]
a_{B-N}, nm	0.1447	0.145	0.1446	0.147	0.151	0.153	0.132

Table 1.
B-N covalent bond length values available in the literature.

$$D_n = \frac{L_C}{\pi} = \frac{a_{B-N}\sqrt{n^2 + nm + m^2}}{\pi}, \tag{4}$$

where $\boldsymbol{a}$ is the length of the unit vector defined through the equilibrium B-N covalent bond length a_{B-N} by $a = \sqrt{3}a_{B-N}$.

There is no agreement among researchers regarding the B-N bond length, therefore different a_{B-N} values can be found in the literature as shown in **Table 1**.

3. Molecular mechanics of 1D and 2D BN nanostructures and equivalent properties of elastic beams

In this study, the NCM/MSM approach was used to evaluate the elastic properties of BNNSs and SWBNNTs. This approach makes use of the equivalence between bonding interactions in the h-BN lattice and the elastic behaviour of beam elements.

As stated by Mayo et al. [66] and Rappé et al. [67], the energy of bonded interactions is defined via those related to bond stretching, U_r, bond bending, U_θ, and bond torsion, U_τ, as follows:

$$U_{bond} = U_r + U_\theta + U_\tau = \frac{1}{2}\Delta k_r(\Delta r)^2 + \frac{1}{2}k_\theta(\Delta\theta)^2 + \frac{1}{2}k_\tau(\Delta\phi)^2, \tag{5}$$

where k_r, k_θ, and k_τ are the bond stretching, bond bending, and torsional resistance force constants, respectively, and Δr, $\Delta\theta$, and $\Delta\phi$ are the bond stretching increment, bond angle bending variation, and angle variation of the twist bond, respectively.

The term of bond torsion energy, U_τ, is a result of merging dihedral angle torsion, U_ϕ, and out-of-plane torsion (also called improper torsion), U_ψ, energy terms, as follows:

$$U_\tau = U_\phi + U_\psi = \frac{1}{2}\left(2k_\phi + k_\psi\right)(\Delta\phi)^2 = \frac{1}{2}k_\tau(\Delta\phi)^2, \tag{6}$$

where k_ϕ and k_ψ are dihedral torsion and inversion force constants, respectively, and $k_\tau = 2k_\phi + k_\psi$.

In its turn, the axial, U_A, bending, U_B, and torsional, U_T, strain energies associated with respective elastic deformations of beams are expressed by:

$$U_A = \frac{1}{2}\frac{E_b A_b}{l}(\Delta l)^2, \tag{7}$$

$$U_B = \frac{1}{2}\frac{E_b I_b}{l}(\Delta\omega)^2, \tag{8}$$

$$U_T = \frac{1}{2}\frac{G_b J_b}{l}(\Delta\vartheta)^2, \tag{9}$$

where E_b and G_b are the Young's and shear moduli of the beam, respectively; l is the beam length; A_b, I_b and J_b are the cross-section area, the moment of inertia, and the polar moment of inertia of the beam, respectively; Δl is the beam axial tensile displacement, ω is the rotational angle at the ends of the beam, $\Delta\vartheta$ is the relative rotation between the ends of the beam.

Consequently, the beam tensile, E_bA_b, bending, E_bI_b, and torsional, G_bJ_b, rigidities can be linked to bond stretching, k_r, bond bending, k_θ, and torsional resistance, k_τ, force constants, respectively, equating, $U_r = U_A$, $U_\theta = U_B$ and $U_\tau = U_T$, from (Eq. (5)) and (Eqs. (7)–(9)), as follows [68]:

$$E_bA_b = k_r l; E_bI_b = k_\theta l; G_bJ_b = k_\tau l. \tag{10}$$

With regard to the k_r, k_θ, and k_τ force constants for the 1D and 2D BN nanostructures, dissimilar values were reported in the literature, depending on the calculation method chosen for this purpose. One of the established methods to calculate force field constants of the diatomic nanostructure is based on Universal Force Fields (UFF) [67] and the other combines *ab initio* DFT calculations with the analytical relationships, coming from molecular mechanics (MM), for the surface Young's modulus, E_s, and the Poisson's ratio, ν, [47]. The E_s and ν values to be replaced in the MM expressions are taken from the literature [64, 69] or obtained from DFT calculations [70]. Tapia et al. [46] obtained the bond force constants from *ab initio* DFT computations, without resorting to MM models. The DREIDING force field (FF) approach [66], which is based solely on the hybridization of atoms, provides the force field constants directly. Apart from the well-known molecular FFs, a classical FF, for describing the B-N bonds, was developed using *ab initio* molecular dynamics (AIMD) simulations combined with lattice dynamics (LD) calculations [65]. The k_r, k_θ, and k_τ force constants for 1D and 2D BN nanostructures, calculated by different approaches are summarised in **Table 2**. It is worth noting that the bond bending force constant, k_θ, depends on the effective charges of the B and N atoms ($Z^*_{1,2}$) and the three-body angles between the bond pairs, B – N – B and N – B – N, which results in two different values $k_{\theta 1}$ and $k_{\theta 2}$, linked by relationship $k_{\theta 1}/k_{\theta 2} = Z_2^{*2}/Z_1^2$ [67].

The values of the bond torsion constant, k_τ, available in the literature are less frequent and with greater discrepancy than those reported for k_r and k_θ force constants. Ansari et al. [70] based their calculation on the link between the bending rigidity of the BN sheet and k_τ, established by molecular mechanics (MM). Genoese et al. [69] calculated the dihedral torsion force constant, k_ϕ, also based on the link between k_ϕ and bending rigidity. Rajan et al. [65] computed two values of the

Reference	Method	k_r, nN/nm	k_θ, nN·nm/rad^2	k_τ, nN·nm/rad^2
Rappé et al. [67]	UFF	676	1.627	—
Li and Chou [48]	DREIDING FF modified	487	0.695	0.625
Rajan et al. [65]	classic FF: AIMD+LD	413	0.383[1] 0.978[2]	1.122[3] 0.428[4]
Jiang and Guo [64]	DFT + MM	595	1.360[1] 0.662[2]	—
Genoese et al. [69]		585	1.347[1] 0.641[2]	0.052[5]
Ansari et al. [70]		620	1.050	2.470
Tapia et al. [46]	DFT	617	0.627	0.132

[1]*$k_{\theta 1}$ for three-body angle of N-B-N.* [2]*$k_{\theta 2}$ for B-N-B.* [3]*$k_{\psi 1}$ for out-of-plane torsion angle of B-N-N-N.* [4]*$k_{\psi 2}$ for N-B-B-B.* [5]*k_ϕ.*

Table 2.
Force field constants for 1D and 2D BN nanostructures reported in the literature.

inversion force constants, $k_{\psi 1}$ and $k_{\psi 2}$ for improper dihedral angles B-N-N-N and N-B-B-B, respectively.

The input parameters for the numerical simulation of 1D and 2D boron nitride nanostructures are calculated based on (Eq. (10)), and assuming the equivalence between the B-N bond length, $a_{\mathrm{B-N}}$, and the beam length, l. The assumption that the beam element has a circular cross-section area, d being its diameter, allows determining the cross-section area, A_b, the moment of inertia, I_b, and the polar moment of inertia, J_b, of the beam, as follows:

$$A_b = \pi d^2/4; I_b = \pi d^4/64; J_b = \pi d^4/32. \quad (11)$$

Regarding **Table 2**, it is possible to conclude that the adequate selection of the values of k_r, k_θ, and k_τ, to calculate the geometrical and elastic properties of beams, which in turn provide input for numerical simulation, is a challenging task. Sakharova et al. [71] studied the influence of input parameters, calculated based on different sets of force field constants, on the elastic properties of BNNTs. It was found that the literature results are in better agreement with the input parameters assessed using values of the force constants close to those reported by Genoese et al. [69] (see **Table 2**). Consequently, in the current study, the bond stretching, k_r, and bond bending, k_θ, force constants from the work of Genoese et al. [69] were chosen to calculate the input parameters for the numerical simulation, as shown in **Table 3**. The torsional resistance constant, k_τ, was calculated by $k_\tau = 2k_\phi + k_\psi$, using $k_\phi = 0.052$ nN nm/rad^2 computed by Genoese et al. [69] and $k_\psi = 0.278$ nN · nm/rad^2 taken from DREIDING FF [66].

4. Geometric characteristics of BN nanosheets and single-walled nanotubes and finite element analysis

Single-layer BNNS with a size of 14.26×13.94 nm^2 was selected for finite element analysis (FEA). With regard of SWBNNTs, non-chiral NTs, zigzag ($\theta = 0°$), and armchair ($\theta = 30°$) were selected for FEA (see **Table 4**). The aspect ratio between

Parameter	Value	Formulation
bond stretching force constant, k_r [69]	585 nN/nm	—
bond bending force constants, $k_{\theta 1}$, $k_{\theta 2}$ [69]	1.347, 0.641 nN·nm/rad^2	—
torsional resistance force constant, k_τ	0.382 nN·nm/rad^2	—
beam length, l, [61]	0.145 nm	$l = a_{B-N}$
diameter, d	0.1649 nm	$d = 2\sqrt{\frac{2(k_{\theta 1}+k_{\theta 2})}{k_r}}$
Young's modulus, E_b	3973 GPa	$E_b = \frac{k_r^2 l}{2\pi(k_{\theta 1}+k_{\theta 2})}$
shear modulus, G_b	764 GPa	$G_b = \frac{k_r^2 k_\tau l}{2\pi(k_{\theta 1}+k_{\theta 2})^2}$
Poisson's ratio, ν_b	0.21	$\nu_b = \frac{k_r l^2 - 3(k_{\theta 1}+k_{\theta 2})}{k_r l^2 + 9(k_{\theta 1}+k_{\theta 2})}$

Table 3.
Input parameters for numerical simulations of BNNSs and SWBNNTs: Geometric and elastic properties of beam elements.

zigzag		armchair	
(n, m)	**D_n, nm**	**(n, m)**	**D_n, nm**
(18, 0)	1.459	(10, 10)	1.404
(20, 0)	1.621	(15, 15)	2.106
(30, 0)	2.431	(20, 20)	2.807
(38, 0)	3.080	(25, 25)	3.509
(43, 0)	3.809	(27, 27)	3.790

Table 4.
Geometric characteristics of the non-chiral SWBNNTs.

nanotube length, L_n, and diameter, D_n, was about 30, thus the NTs elastic response is independent of L_n.

The FE meshes of the BNNS and SWBNNTs, used in FEA, were obtained with the help of the Nanotube Modeller© software. This software produces the Program Database files, which are then converted, with the *InterfaceNanotubes.NM* in-house application [71], to the appropriate format to be used by the ABAQUS® code.

The mechanical behaviour of the BN nanosheet was studied under numerical tensile and in-plane shear tests, using the ABAQUS® FE code. **Figure 3** shows the geometry of the NS and the boundary conditions of the three studied loading cases.

In the first loading case, the nodes on the left edge of the BNNS were fixed and an axial tensile force, P_x, was applied on the opposite (right) edge (**Figure 3a**). In the second case, the nodes on the bottom edge of the BNNS were fixed, while an axial transverse force, P_y, was applied on the top edge of the nanosheet (**Figure 3b**). In the third loading case, the boundary conditions were the same as in the second case, and a shear force, N_x, was applied to the upper edge nodes of the BNNs (**Figure 3c**).

Under the applied force P_x, the BNNS elongates in the x- direction, which leads to axial displacement, u_x. The Young's modulus along the x-axis, E_x, is determined as follows [46]:

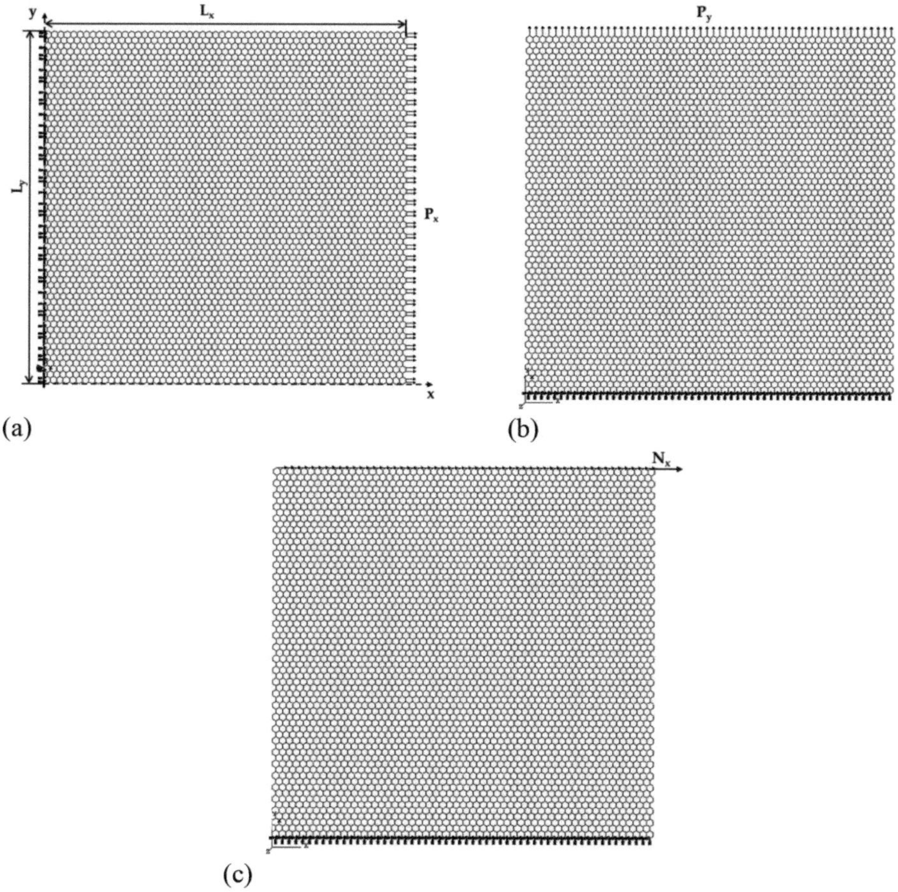

Figure 3.
Schematic representation of: (a) tensile loading in the x - direction (zigzag configuration); (b) tensile loading in the y- direction (armchair configuration); (c) in-plane shear loading in the x-direction for the BNNSs. The geometrical parameters and the boundary conditions of the nanosheet are also presented.

$$E_x = \frac{P_x L_x}{u_x L_y t_n}, \tag{12}$$

where L_x and L_y are the BNNS side lengths as shown in **Figure 3**; t_n is the nanosheet thickness.

The Young's modulus along the y-axis, E_y, is calculated by the following expression [46]:

$$E_y = \frac{P_y L_y}{v_y L_x t_n}, \tag{13}$$

where P_y is applied transverse force, v_y is the displacement of the BNNS in the y-direction, taken from the FEA.

To calculate the shear strain, γ_{xy}, the displacement of the BNNS in the x-direction under the in-plane shear force (N_x), v_x, was taken from FEA. Consequently, the shear modulus, G_{xy}, of the BNNS is calculated by the following expression [46]:

Reference	t_n, nm	Reference	t_n, nm
Kudin et al. [74]	0.0936; 0.335	Zhang et al. [75]	0.314
Vijayaraghavan and Zhang [76]	0.105	Zhao and Xue [41]	0.330
Tapia et al. [46]	0.106	Yan et al. [52]	0.333

Table 5.
Nanotube wall thickness values, t_n, reported in the literature.

$$G_{xy} = \frac{N_x}{\gamma_{xy} L_x t_n}, \gamma_{xy} = \tan \frac{v_x}{L_y}. \tag{14}$$

where L_x and L_y are the NS side lengths, t_n is the NS thickness.

The ABAQUS® FE code was also used to evaluate the elastic response of SWBNNTs under tensile, bending, and torsion loading. The boundary conditions consisted in fixing the edge nodes at one end of the NT. The axial tensile force, F_a, the transverse force, F_t, and the torsional moment, M_T, were applied to the other end of the NT, to carry out tensile, bending, and torsion tests. In the last case, the edge nodes cannot move in the radial direction.

The axial displacement, u_a, the transverse displacement, u_t, and the twist angle, ω, were obtained from the FEA of the respective test. This allows calculating the tensile, EA, bending, EI, and torsional, GJ, rigidities of the non-chiral SWBNNTs with length L_n as follows:

$$EA = \frac{F_a L_n}{u_a}, EI = \frac{F_t L_n^3}{u_t^3}, GJ = \frac{M_T}{\omega}. \tag{15}$$

Making use of (Eq. (15)), the Young's, E_{NT}, and shear, G_{NT}, moduli of the SWBNNTs are determined by the following expressions [72]:

$$E_{NT} = \frac{EA}{\pi t_n \sqrt{8\left(\frac{EI}{EA}\right) - t_n^2}}, \tag{16}$$

$$G_{NT} = \frac{GJ}{2\pi\left(\frac{EI}{EA}\right) t_n \sqrt{8\left(\frac{EI}{EA}\right) - t_n^2}}, \tag{17}$$

where t_n is the nanotube wall thickness, the same parameter as the nanosheet thickness in (Eqs. (12)–(14)).

In the present study, the wall thickness value of the SWBNNTs was considered equal to the graphite interlayer spacing, $t_n = 0.34$ nm. Such value was confirmed experimentally by TEM, $t_n = 0.338 \pm 0.004$ nm [73], and commonly used by the research community [43, 48, 49, 71]. Still, there is no agreement on the t_n value as shown in **Table 5**.

5. Elastic moduli of the boron nitride nanosheets (BNNS)

Figure 4a shows Young's modulus of the BNNS, along the x-axis (zigzag direction), E_x, and the y-axis (armchair direction), E_y, evaluated respectively by (Eqs. (12)

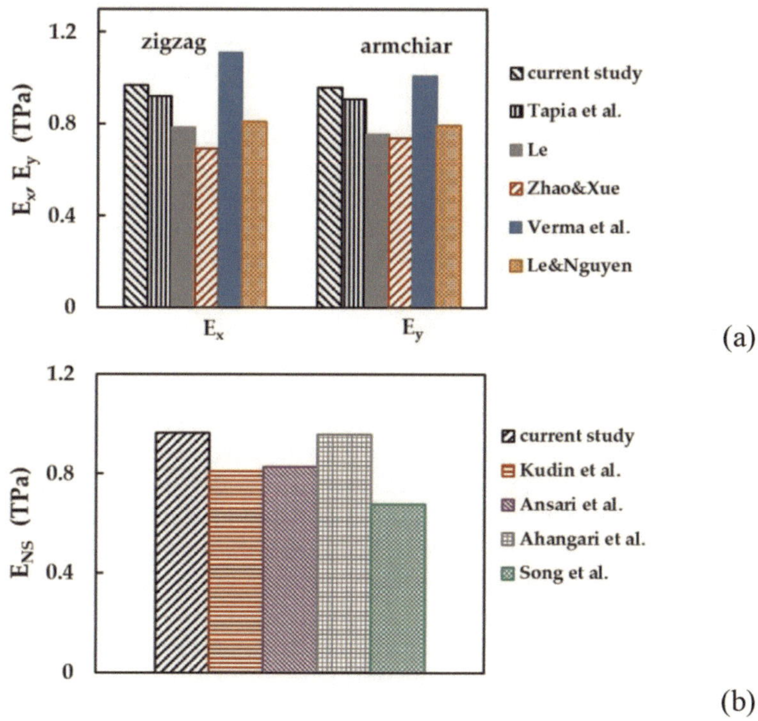

Figure 4.
(a) Current Young's moduli, E_x (zigzag) and E_y (armchair), of BNNSs and those by Tapia et al. [46], Le [40], Zhao and Xue [41], Verma et al. [77], and Le and Nguyen [78]; (b) current Young's modulus, E_{NS}, of BNNS and those by Kudin et al. [74], Ansari et al. [49], Ahangari et al. [39], and Song et al. [1].

and (13)), as well as E_x and E_y results available in the literature. For cases in which only one Young's modulus was reported, the comparison was carried out with the average value $E_{NS} = (E_x + E_y)/2$, as shown in **Figure 4b**. The value of E_{NS}, measured by the nanoindentation test, performed in AFM, in the work by Song et al. [1], is also shown in **Figure 4b**. Since Song et al. [1] reported the surface Young's modulus (the product of Young's modulus by the nanosheet thickness, $E_{sNS} = E_{NS}t_n$), E_{NS} was evaluated using t_n = 0.33 nm, as suggested in their study.

The scatter of Young's moduli results, E_x and E_y, available in the literature is evidenced in **Figure 4a**. Good agreement was found between the current E_x (zigzag direction) and E_y (armchair direction) values and those evaluated by Tapia et al. [46], with a difference of 5.4 and 5.7% for E_x and E_y, respectively. The work by Tapia et al. [46] shares the same modelling (NCN/MSM with beams) and calculation approach as the present study. The values E_x and E_y reported by Verma et al. [77] in their MD study employing TB potential, are 14.5 and 4.2% higher, respectively, than those obtained in the present study. For Young's moduli along the x-axis and y-axis calculated by Le [40], differences of 23.6 and 26.8% are seen when compared with the current E_x and E_y moduli, respectively. Le and Nguyen [78] reported E_x and E_y values of 19.8 and 20.8% lower than those evaluated by (Eqs. (12) and (13)), respectively. Despite the similar values of E_x and E_y obtained in the works by Le [40] and Le and Nguyen [78], these studies used different modelling methods, which are MD simulations with Tersoff and Tersoff-like potentials and FE simulations under NCM/MSM approach based on harmonic force fields, respectively. The largest difference of 40.0% (for E_x) and 29.8% (for E_y) were found between current Young's modulus values and those by Zhao and Xue [41], who employed MD simulations with Tersoff potential.

Particularly good agreement (difference of 0.8%) is observed between the current Young's modulus of the BNNS, E_{NS}, and that assessed by Ahangari et al. [39] using ab initio DFT calculations (see, **Figure 4b**). The BNNS Young's moduli reported by Kudin et al. [74] and Ansari et al. [49] are 19.0 and 16.2% smaller, respectively, than E_{NS} calculated in the present study. In the work of Kudin et al. [74] Young's modulus was evaluated using *ab initio* calculations. Ansari et al. [49] employed an analytical solution in the framework of the NCM/MSM approach for this purpose. The biggest difference of about 43% is seen when the current BNNS Young's modulus is compared with that obtained by nanoindentation experiments [1]. This comparatively lower value of E_{NS}, evaluated in a study by Song et al. [1], can be related to the presence of intrinsic defects and their distribution in boron nitride NS.

The results presented in **Figure 4a** indicate that the BN nanosheets are not transversely isotropic, i.e. Young's moduli in the zigzag and armchair directions are not equal to each other, $E_x \neq E_y$. The present study, as well as those by Le [40], Zhao and Xue [41], Tapia et al. [46], Verma et al. [77], and Le and Nguyen [78], suggest a mild anisotropy of BNNSs. Such anisotropic nanosheet behaviour can be quantified by the ratio between Young's moduli for zigzag and armchair configurations, E_x/E_y. In **Figure 5**, the current ratio E_x/E_y is compared with those calculated from the results available in the literature.

The current ratio of $E_x/E_y \approx 1.01$ is in very good agreement with $E_x/E_y \approx 1.01$, 1.04, 1.02 found in the works by Tapia et al. [46], Le [40], and Le and Nguyen [78], respectively. In contrast to other results under analysis, Zhao and Xue [41] reported Young's modulus for the zigzag NS configuration smaller than for armchair, with the ratio $E_x/E_y \approx 0.94$. Verma et al. [77] indicated the highest anisotropy ratio, among others presented in **Figure 5**, $E_x/E_y \approx 1.10$. The BN nanosheet anisotropy can possibly be explained by the alignment of the bonds in relation to the loading path, which causes it to lose crystalline symmetry, as there are two types of atoms in each structure.

The evaluation of the BNNS shear modulus has practically not been the research focus so far. **Figure 6** shows the shear moduli of the BNNS, G_{xy}, evaluated by

Figure 5.
Comparison of the current E_x/E_y ratio with those obtained from the literature by Tapia et al. [46], Le [40], Zhao and Xue [41], Verma et al. [77], and Le and Nguyen [78].

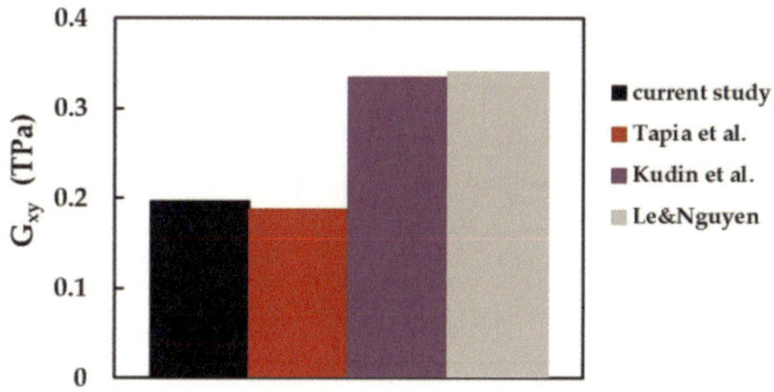

Figure 6.
Current shear modulus, G_{xy} of BNNSs and those by Tapia et al. [46], Kudin et al. [74], and Le and Nguyen [78].

Reference	Method	t_n, nm	E_x, TPa	E_y, TPa	E_x/E_y	G_{xy}, TPa	NS size, nm^2
Kudin et al. [74]	*ab initio*	0.335	0.810		—	0.334	—
Verma et al. [77]	MD: TB potential	0.333	1.110	1.010	1.10	—	8.67 × 10.22
Zhao and Xue [41]	MD: Tersoff potential	0.330	0.692	0.739	0.94	—	12.00 × 12.00
Le [40]	MD: Tersoff and Tersoff-like potentials	0.335	0.786	0.756	1.04	—	10.36 × 10.22
Ahangari et al. [39]	*ab initio* DFT	0.320	0.956		—	—	1.005 × 1.132
Le and Nguyen [78]	NCM/MSM: harmonic force fields	0.335	0.809	0.794	1.02	0.340	10.42 × 10.29
Ansari et al. [49]	NCM/MSM: analytical solution	0.340	0.829		—	—	—
Tapia et al. [46]	NCM/MSM: beams	0.106	0.920	0.908	1.01	0.188	14.79 × 14.71
Song et al. [1]	Experimental: nanoindentation+AFM	0.330	0.676		—	—	7.85 × 10^4 (circular)
current study	NCM/MSM: beams	0.340	0.969	0.959	1.01	0.196	14.26 × 13.94

Table 6.
Current results of Young's and shear moduli for boron nitride nanosheets and those reported in the literature.

(Eq. (14)), with the G_{xy} results available in the literature. The value of G_{xy} evaluated in the current work is in good concordance (the difference of 4.6%) with that reported by Tapia et al. [46]. On the other hand, the current shear modulus is more than 40% smaller than G_{xy} evaluated by Kudin et al. [74] and Le and Nguyen [78].

To facilitate understanding, the results of the elastic moduli shown in **Figures 4–6**, are summarised in **Table 6**.

6. Elastic moduli of the single-walled boron nitride nanotubes (SWBNNTs)

Figure 7a shows Young's modulus, E_{NT}, of the non-chiral, zigzag, and armchair SWBNNTs, as a function of the NT diameter, D_n. Young's modulus was evaluated by (Eq. (16)), which makes use of the tensile, EA, and bending, EI, rigidities calculated

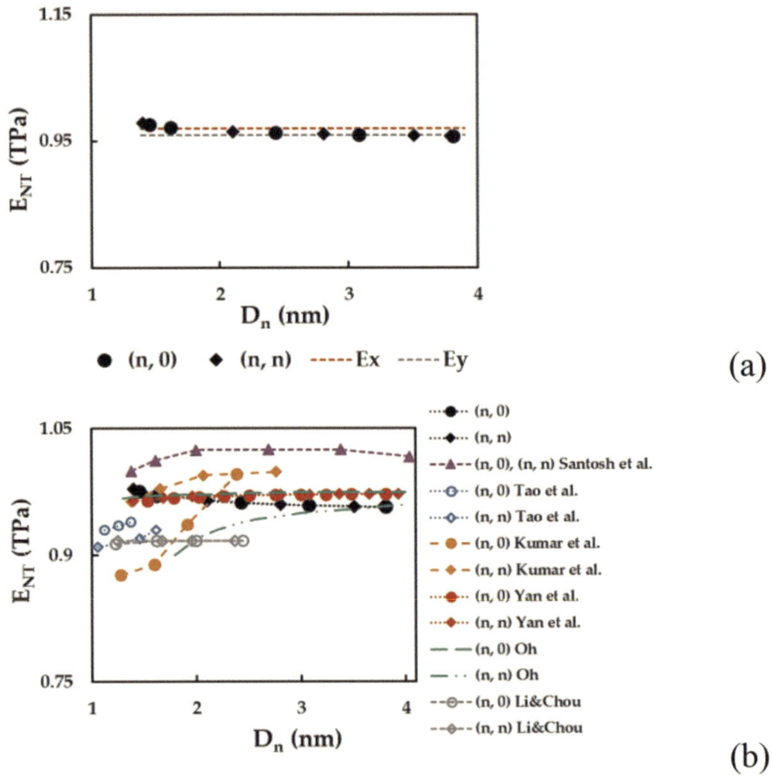

Figure 7.
(a) Evolution of Young's modulus, E_{NT}, with NT diameter, D_n, for non-chiral SWBNNTs; (b) comparison of the current results of Young's modulus of SWBNNTs with those reported in the literature [42–45, 48, 52].

with (Eq. (15)) based on the respective numerical test results. Young's moduli of boron nitride NS in zigzag, E_x, and armchair, E_y, directions are also plotted in **Figure 7a**. **Figure 7b** compares the values of E_{NT} obtained in the present study with those reported in the literature, selected so that different modelling and calculation approaches were taken into consideration. The Young's moduli of non-chiral boron nitride NTs with diameters range similar to the one studied in the present work were chosen for comparison purpose.

The E_{NT} evolutions with the nanotube diameter for zigzag and armchair boron nitride NTs coincide, which establishes a unique trend for both $(n, 0)$ and (n, n) non-chiral NTs (see, **Figure 7a**). The Young's modulus of the SWBNNTs, after a slight decrease at the beginning of the trend, converges to a nearly constant value, $E_{NT} \approx$ 0.958 TPa, with increasing of D_n. This stabilised value, on the one hand, tends to Young's modulus of BNNS in the zigzag direction, E_x, and, on the other hand, is equal to the E_y value in the armchair direction.

It can be concluded from **Figure 7b** that the current SWBNNTs Young's modulus results are in reasonably good consonance with those reported in the literature. Particularly good agreement (difference of 0.05%) was found between the current value of E_{NT} and that evaluated for (n, n) NTs with $D_n > 3.0\ nm$ by Oh [45] in a study employing the CM approach. The difference becomes 1.73% when compared with the E_{NT} value for (n, n) NTs in the entire range of diameters considered, obtained in the same work [45]. For Young's moduli calculated by Yan et al. [52], based on the analysis of longitudinal vibrations under the NCM/MSM approach, and by Tao et al. [43], employing MD combined with the FE model, the differences for the current value of E_{NT} are about 1.5 and 2.5%, respectively. The greatest differences of about 3.9, 4.5, and 5.8% are observed when E_{NT} evaluated by (Eq. (16)) is compared with Young's modulus reported by Kumar et al. [44], Li and Chou [48], and Santosh et al. [42], respectively. The works of Santosh et al. [42] and Kumar et al. [44] both used MD simulation, with the difference in the way of describing the interatomic interactions in the h-BN lattice, force-constant method, and REBO potential, respectively. Li and Chou [48] investigated the elastic behaviour of the SWBNNTs within the framework of the NCM/MSM approach with beam elements to model the B-N bond, as in the present study, using only the other force field constants (see, **Table 2**).

It is worth noting that Young's modulus values evaluated in the present study for the BNNS and SWBNNTs are close to those typically obtained for the CNTs ($E_{CNT} \approx 1.0$ TPa), which indicates that the 1D and 2D BN nanostructures are appropriate candidates to replace carbon counterparts in various technological applications or to envisage the design of hybrid structures, constituted by NSs or NTs of carbon and boron-nitride.

Figure 8a shows the shear modulus, G_{NT}, of the SWBNNTs, as a function of the NT diameter, D_n. The shear modulus was evaluated by (Eq. (17)), using tensile, EA, bending, EI, and torsional, GJ, rigidities derived by (Eq. (15)) from FEA results.

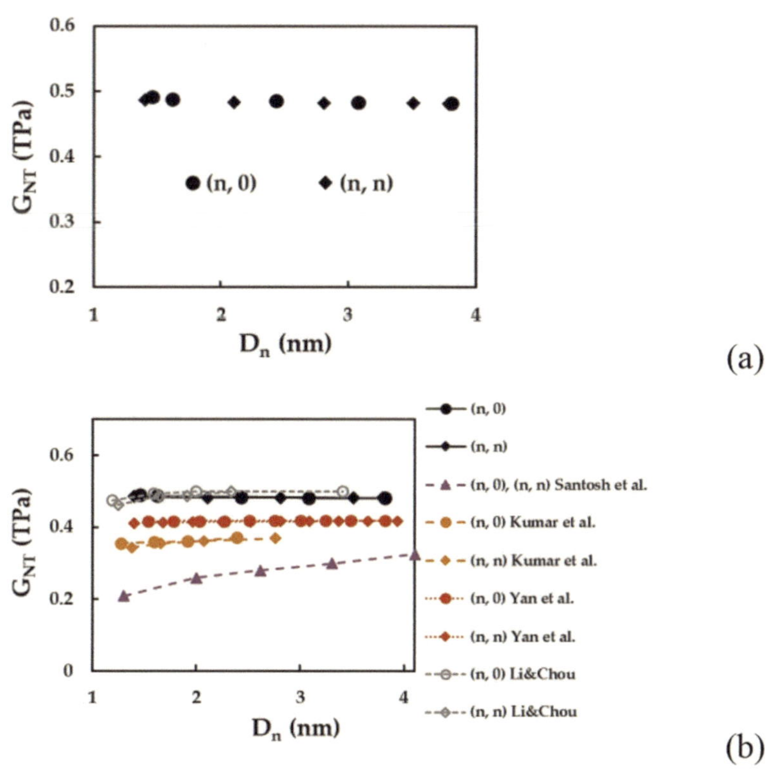

Figure 8.
(a) Evolutions of the current shear modulus, G_{NT}, with NT diameter, D_n, for the non-chiral SWBNNTs; (b) comparison of the current results of the shear modulus of SWBNNTs with those reported in the literature [42, 44, 48, 52].

Figure 8b shows the current shear modulus values with those available in the literature, for comparison purposes.

The shear modulus, G_{NT}, of the SWBNNTs is almost constant with increasing NT diameter, D_n, except at the beginning of the evolution $G_{NT} = f(D_n)$, where the insignificant decrease of the G_{NT} is observed (see, **Figure 8a**). Regarding the results available in the literature, a reasonable agreement, with a difference ≈ 3.6%, is observed when the current values of G_{NT} are compared to those evaluated by Li and Chou [48], who employed the same modelling approach as in the present study but using a different set of force field constants (see, **Figure 8b**). However, greater differences occur for the other available G_{NT} values. As previously discussed [71], the

Reference	Method	t_n, nm	NT type	E_{NT}, TPA	G_{NT}, TPA	Comments
Santosh et al. [42]	MD: force constant approach	0.340	(n, 0), (n, n)	1.017	0.326	converged average value
Tao et al. [43]	MD: TB potential + FE model		(n, 0) (n, n)	0.930 0.911	—	
Kumar et al. [44]	MD: REBO potential	0.330	(n, 0) (n, n)	0.967 0.997	0.364 0.366	
Oh [45]	CM: continuum lattice thermodynamic + TB potential		(n, 0) (n, n)	0.975 0.960	—	
Yan et al. [52]	NCM/MSM: longitudinal and torsional vibrations	0.333	(n, 0), (n, n)	0.972	0.418	
Li and Chou [48]	NCM/MSM: beam elements	0.340	(n, 0) (n, n)	0.913 0.916	0.475 0.465	
Chopra and Zettl [55]	Experimental: TEM (thermal vibrations)	—	MW	1.22 ±0.24	—	—
Suryavanshi et al. [56]	Experimental: TEM (electric-field-induced resonance)	—	MW	0.722 ±0.14	—	average value for 18 NTs
Arenal et al. [53]	Experimental: HRTEM-AFM	0.070 0.340	SW	1.11 ±0.17 0.25 ±0.04	—	—
Goldberg et al. [58]	Experimental: AFM – TEM (bending test)	—	MW	0.5-0.6	—	average value
Ghassemi et al. [59]		—	MW	0.50 ±0.10	—	average value for 5 NTs
Tanur et al. [60]	Experimental: AFM (three-point bending)	—	MW	0.760 ±0.03	0.07 ±0.01	average value for 20 NTs
Zhou et al. [57]	Experimental: HRTEM (high-order resonance)	—	MW	0.906	—	average value
Chen et al. [54]	Experimental: TEM (compression circles)	—	MW	1.050–1.370	—	—
Current study	NCM/MSM: beam elements	0.34	(n, 0), (n, n)	0.958	0.480	converged average value

Table 7.
Comparison of current Young's and shear moduli results for BNNTs with those reported in the literature.

SWBNNTs shear modulus results are relatively uncommon in the literature so far, in addition to showing a significant scattering of reported values.

To facilitate the understanding of the results of the SWBNNTs elastic moduli presented in **Figures** 7 and **8**, they are summarised in **Table** 7, which also includes the experimental results from the literature.

It is possible to conclude from this table that the current Young's modulus results are in reasonable agreement with most of the experimental values of E_{NT}. With regard to the shear modulus, as far as we know, only Tanur et al. [60] reported a very low G_{NT} value for the case of MWBNNTs, which was explained by the shear effects occurring between adjacent layers.

7. Conclusions

In the present numerical simulation study, Young's and shear moduli of 1D (nanotubes) and 2D (nanosheets) boron nitride structures were evaluated based on the NCM/MSM approach.

The current result of the elastic moduli points to a mild anisotropy of the boron nitride nanosheets. The Young's modulus of the BNNS in the zigzag direction, E_x, is greater than that in the armchair direction, E_y, with the anisotropy ratio $E_x/E_y \approx 1.01$, whose value is in good concordance with those reported in the literature.

The Young's and shear moduli of single-walled boron nitride nanotubes become quasi-constant with increasing nanotube diameter, D_n, after a slight decrease at the beginning of the trend. This quasi-constant value of the SWBNNTs Young's modulus tends to the BNNS Young's modulus in a zigzag direction, E_x, and coincides with that in the armchair direction, E_y.

The current results establish a reference for the evaluation of the elastic properties of the boron nitride nanosheets and nanotubes by numerical methods.

Acknowledgements

This research is sponsored by FEDER funds through the program COMPETE—Programa Operacional Factores de Competitividade—and by national funds through FCT, Fundação para a Ciência e a Tecnologia, under the projects CEMMPRE—UIDB/00285/2020 and ARISE—LA/P/0112/2020

Conflict of interest

The authors declare no conflict of interest.

Author details

Nataliya A. Sakharova[1*], Jorge M. Antunes[1,2], André F.G. Pereira[1], Bruno M. Chaparro[2] and José V. Fernandes[1]

1 Centre for Mechanical Engineering, Materials and Processes (CEMMPRE) - Advanced Production and Intelligent Systems, Associated Laboratory (ARISE), University of Coimbra, Coimbra, Portugal

2 Abrantes High School of Technology, Polytechnic Institute of Tomar, Tomar, Portugal

*Address all correspondence to: nataliya.sakharova@dem.uc.pt

References

[1] Song L, Ci L, Lu H, Sorokin PB, Jin C, Ni J, et al. Large scale growth and characterization of atomic hexagonal boron nitride layers. Nano Letters. 2010; **10**(8):3209-3215. DOI: 10.1021/nl1022139

[2] Suzuki S, Hibino H. Chemical vapor deposition of hexagonal boron nitride. e-Journal of Surface Science and Nanotechnology. 2012;**10**:133-138. DOI: 10.1380/ejssnt.2012.133

[3] Dean CR, Young AF, Meric I, Lee C, Wang L, Sorgenfrei S, et al. Boron nitride substrates for high-quality graphene electronics. Nature Nanotechnology. 2010;**5**(10):722-726. DOI: 10.1038/nnano.2010.172

[4] Liu Z, Ma L, Shi G, Zhou W, Gong Y, Lei S, et al. In-plane heterostructures of graphene and hexagonal boron nitride with controlled domain sizes. Nature Nanotechnology. 2013;**8**(2):119-124. DOI: 10.1038/nnano.2012.256

[5] Wang Y, Zhou V, Xie Y, Zheng X-Q, Feng PX-L. Optical contrast signatures of hexagonal boron nitride on a device platform. Optical Materials Express. 2019;**9**(3):1223-1232. DOI: 10.1364/OME.9.001223

[6] Liu Y, Weiss NO, Duan X, Cheng H-C, Huang Y, Duan X. Van der Waals heterostructures and devices. Nature Reviews Materials. 2016;**1**(9):16042. DOI: 10.1038/natrevmats.2016.42

[7] Zheng X-Q, Lee J, Feng PX-L. Hexagonal boron nitride nanomechanical resonators with spatially visualized motion. Microsystems & Nanoengineering. 2017;**3**:17038. DOI: 10.1038/micronano.2017.38

[8] Zhang Y, Shi Q, Liu Y, Wang Y, Meng Z, Xiao C, et al. Hexagonal boron nitride with designed nanopores as a high-efficiency membrane for separating gaseous hydrogen from methane. Journal of Physical Chemistry C. 2015;**119**(34): 19826-19831. DOI: 10.1021/acs.jpcc.5b04918

[9] Gao H, Shi Q, Rao D, Zhang Y, Su J, Liu Y, et al. Rational design and strain engineering of nanoporous boron nitride nanosheet membranes for water desalination. Journal of Physical Chemistry C. 2017;**121**(40):22105-22113. DOI: 10.1021/acs.jpcc.7b06480

[10] Tran TT, Bray K, Ford MJ, Toth M, Aharonovich I. Quantum emission from hexagonal boron nitride monolayers. Nature Nanotechnology. 2016;**11**(1): 37-41. DOI: 10.1038/nnano.2015.242

[11] Khan AA, Ahmad R, Ahmad I. Silicon carbide and III-nitrides nanosheets: Promising anodes for Mg-ion batteries. Materials Chemistry and Physics. 2021;**257**:123785. DOI: 10.1016/j.matchemphys.2020.123785

[12] Novoselov KS, Jiang D, Schedin F, Booth TJ, Khotkevich VV, Morozov SV, et al. Two-dimensional atomic crystals. PNAS. 2005;**102**(30):10451-10453. DOI: 10.1073/pnas.0502848102

[13] Tay RY, Tsang SH, Loeblein M, Chow WL, Loh GC, Toh JW, et al. Direct growth of nanocrystalline hexagonal boron nitride films on dielectric substrates. Applied Physics Letters. 2015; **106**(10):101901. DOI: 10.1063/1.4914474

[14] Mirzaee M, Rashidi A, Zolriasatein A, Abadchi MR. A simple, low cost, and template-free method for synthesis of boron nitride using different precursors. Ceramics International. 2021;**47**(5):5977-5984. DOI: 10.1016/j.ceramint.2020.10.171

[15] Pacilé D, Meyer JC, Girit ÇO, Zettl A. The two-dimensional phase of boron nitride: Few-atomic layer sheets and suspended membranes. Applied Physics Letters. 2008;**92**:133107. DOI: 10.1063/1.2903702

[16] Yanar N, Yang E, Park H, Son M, Choi H. Boron nitride nanotube (BNNT) membranes for energy and environmental applications. Membranes. 2020;**10**(20):430. DOI: 10.3390/membranes10120430

[17] Hilder TA, Gordon D, Chung S-H. Salt rejection and water transport through boron nitride nanotubes. Small. 2009;**5**:2183-2190. DOI: 10.1002/smll.200900349

[18] Siria A, Poncharal P, Biance A-L, Fulcrand R, Blase X, Purcell ST, et al. Giant osmotic energy conversion measured in a single transmembrane boron nitride nanotube. Nature. 2013; **494**:455-458. DOI: 10.1038/nature11876

[19] Roudbari MA, Ansari R. Single-walled boron nitride nanotube as nano-sensor. Continuum Mechanics and Thermodynamics. 2020;**32**:729-748. DOI: 10.1007/s00161-018-0719-6

[20] Chowdhury R, Adhikari S. Boron-nitride nanotubes as zeptogram-scale bionanosensors: Theoretical investigations. IEEE Transactions on Nanotechnology. 2011;**10**(4):659-667. DOI: 10.1109/TNANO.2010.2060492

[21] Chen Y, Fitz Gerald J, Williams JS, Bulcock S. Synthesis of boron nitride nanotubes at low temperatures using reactive ball milling. Chemical Physics Letters. 1999;**299**(3):260-264. DOI: 10.1016/S0009-2614(98)01252-4

[22] Kim J, Lee S, Uhm YR, Jun J, Rhee CK, Kim GM. Synthesis and growth of boron nitride nanotubes by a ball milling–annealing process. Acta Materialia. 2011;**59**(7):2807-2813. DOI: 10.1016/j.actamat.2011.01.019

[23] Yu J, Chen Y, Wuhrer R, Liu Z, Ringer SP. In situ formation of BN nanotubes during nitriding reactions. Chemistry of Materials. 2005;**17**(20): 5172-5176. DOI: 10.1021/cm050966f

[24] Ahmad P, Khandaker MU, Khana ZR, Amina YM. Synthesis of boron nitride nanotubes via chemical vapour deposition: A comprehensive review. RSC Advances. 2015;**5**(44): 35116-35137. DOI: 10.1039/C5RA01594D

[25] Lourie OR, Jones CR, Bartlett BM, Gibbons PC, Ruoff RS, Buhro WE. CVD growth of boron nitride nanotubes. Chemistry of Materials. 2000;**12**(7): 1808-1810. DOI: 10.1021/cm000157q

[26] Kim MJ, Chatterjee S, Kim SM, Stach EA, Bradley MG, Pender MJ, et al. Double-walled boron nitride nanotubes grown by floating catalyst chemical vapor deposition. Nano Letters. 2008; **8**(10):3298-3302. DOI: 10.1021/nl8016835

[27] Huang Y, Lin J, Tang C, Bando Y, Zhi C, Zhai T, et al. Bulk synthesis, growth mechanism and properties of highly pure ultrafine boron nitride nanotubes with diameters of sub-10 nm. Nanotechnol. 2011;**22**(14):145602. DOI: 10.1088/0957-4484/22/14/145602

[28] Golberg D, Bando Y, Eremets M, Takemura K, Kurashima K, Yusa H. Nanotubes in boron nitride laser heated at high pressure. Applied Physics Letters. 1996;**69**(14):2045-2047. DOI: 10.1063/1.116874

[29] Arenal R, Stephan O, Cochon J-L, Loiseau A. Root-growth mechanism for single-walled boron nitride nanotubes in laser vaporization technique. Journal of

the American Chemical Society. 2007; **129**(51):16183-16189. DOI: 10.1021/ja076135n

[30] Cau M, Dorval N, Attal-Trétout B, Cochon JL, Cao B, Bresson L, et al. LASER-based diagnostics applied to the study of BN nanotubes synthesis. Journal of Nanoscience and Nanotechnology. 2008;**8**(11):6129-6140. DOI: 10.1166/jnn.2008.sw14

[31] Smith MW, Jordan KC, Park C, Kim JW, Lillehei PT, Crooks R, et al. Very long single-and few-walled boron nitride nanotubes via the pressurized vapor/condenser method. Nanotechnology. 2009;**20**(50):505604. DOI: 10.1088/0957-4484/20/50/505604

[32] Gnoffo PA, Fay CC. Laser vaporization and plume chemistry in a boron nitride nanotube production rig. Journal of Thermophysics and Heat Transfer. 2013;**27**(3):369-381. DOI: 10.2514/1.T3996

[33] Schutze A, Jeong JY, Babayan SE, Park J, Selwyn GS, Hicks RF. The atmospheric-pressure plasma jet: A review and comparison to other plasma sources. IEEE Transactions on Plasma Science. 1998;**26**(6):1685-1694. DOI: 10.1109/27.747887

[34] Kim KS, Kingston CT, Hrdina A, Jakubinek MB, Guan J, Plunkett M, et al. Hydrogen-catalyzed, pilot-scale production of small-diameter boron nitride nanotubes and their macroscopic assemblies. ACS Nano. 2014;**8**(6): 6211-6220. DOI: 10.1021/nn501661p

[35] Kim KS, Couillard M, Shin H, Plunkett M, Ruth D, Kingston CT, et al. Role of hydrogen in high-yield growth of boron nitride nanotubes at atmospheric pressure by induction thermal plasma. ACS Nano. 2018;**12**(1):884-893. DOI: 10.1021/acsnano.7b08708

[36] Fathalizadeh A, Pham T, Mickelson W, Zettl A. Scaled synthesis of boron nitride nanotubes, nanoribbons, and nanococoons using direct feedstock injection into an extended-pressure, inductively-coupled thermal plasma. Nano Letters. 2014; **14**(8):4881-4886. DOI: 10.1021/nl5022915

[37] Huang Z, Lü T-Y, Wang H-Q, Yang S-W, Zheng J-C. Electronic and thermoelectric properties of the group-III nitrides (BN, AlN and GaN) atomic sheets under biaxial strains. Computational Materials Science. 2017; **130**:232-241. DOI: 10.1016/j.commatsci.2017.01.013

[38] Amorim B, Cortijo A, de Juan F, Grushin AG, Guinea F, Gutiérrez-Rubio A, et al. Novel effects of strains in graphene and other two dimensional materials. Physics Reports. 2016;**617**: 1-54. DOI: 10.1016/j.physrep.2015.12.006

[39] Ahangari MG, Fereidoon A, Mashhadzadeh AH. Interlayer interaction and mechanical properties in multi-layer graphene, boron-nitride, Aluminum-nitride and gallium-nitride graphene-like structure: A quantum-mechanical DFT study. Superlattices and Microstructures. 2017;**112**:30-45. DOI: 10.1016/j.spmi.2017.09.005

[40] Le MQ. Atomistic study on the tensile properties of hexagonal AlN, BN, GaN, InN and SiC sheets. Journal of Computational and Theoretical Nanoscience. 2014;**11**(6):1458-1464. DOI: 10.1166/jctn.2014.3518

[41] Zhao S, Xue J. Mechanical properties of hybrid graphene and hexagonal boron nitride sheets as revealed by molecular dynamic simulations. Journal of Physics D: Applied Physics. 2013;**46**:135303. DOI: 10.1088/0022-3727/46/13/135303

[42] Santosh M, Maiti PK, Sood AK. Elastic properties of boron nitride nanotubes and their comparison with carbon nanotubes. Journal of Nanoscience and Nanotechnology. 2009; **9**(9):5425-5430. DOI: 10.1166/jnn.2009.1197

[43] Tao J, Xu G, Sun Y. Elastic properties of boron-nitride nanotubes through an atomic simulation method. Mathematical Problems in Engineering. 2015;**2015**: 240547. DOI: 10.1155/2015/240547

[44] Kumar D, Verma V, Dharamvir K, Bhatti HS. Elastic moduli of boron nitride, aluminium nitride and gallium nitride nanotubes using second generation reactive empirical bond order potential. Multidiscipline Modeling in Materials and Structures. 2015;**11**(1): 2-15. DOI: 10.1108/MMMS-01-2014-0006

[45] Oh E-S. Elastic properties of boron-nitride nanotubes through the continuum lattice approach. Materials Letters. 2010;**64**(7):859-862. DOI: 10.1016/j.matlet.2010.01.041

[46] Tapia A, Cab C, Hernández-Pérez A, Villanueva C, Peñuñuri F, Avilés F. The bond force constants and elastic properties of boron nitride nanosheets and nanoribbons using a hierarchical modeling approach. Physica E 2017;89: 183-193DOI: 10.1016/j.physe.2016.12.003

[47] Genoese A, Genoese A, Salerno G. On the nanoscale behaviour of single-wall C, BN and SiC nanotubes. Acta Mechanica. 2019;**230**:1105-1128. DOI: 10.1007/s00707-018-2336-7

[48] Li C, Chou T-W. Static and dynamic properties of single-walled boron nitride nanotubes. Journal of Nanoscience and Nanotechnology. 2006;**6**(1):54-60. DOI: 10.1166/jnn.2006.063

[49] Ansari R, Mirnezhad M, Sahmani S. Prediction of chirality- and size-dependent elastic properties of single-walled boron nitride nanotubes based on an accurate molecular mechanics model. Superlattices and Microstructures. 2015; **80**:196-205. DOI: 10.1016/j.spmi.2014.12.033

[50] Zakaria AZ. A two-section beam element to model the B-N covalent bonds in boron nitride nanotubes. Materials Research Bulletin. 2022;**145**: 111533. DOI: 10.1016/j.materresbull.2021.111533

[51] Giannopoulos GI, Kontoni D-PN, Georgantzinos SK. Efficient FEM simulation of static and free vibration behaviour of single walled boron nitride nanotubes. Superlattices and Microstructures. 2016;**96**:111-120. DOI: 10.1016/j.spmi.2016.05.016

[52] Yan JW, He JB, Tong LH. Longitudinal and torsional vibration characteristics of boron nitride nanotubes. Journal of Vibration Engineering & Technologies. 2019;7: 205-215. DOI: 10.1007/s42417-019-00113-4

[53] Arenal R, Wang MS, Xu Z, Loiseau A, Golberg D. Young modulus, mechanical and electrical properties of isolated individual and bundled single-walled boron nitride nanotubes. Nanotechnology. 2011;**22**(26):265704. DOI: 10.1088/0957-4484/22/26/265704

[54] Chen G, Lu H, Cui J, Yu H, Wang B, Liu Y, et al. In situ real-time study buckling behavior of boron nitride nanotubes with axial compression by TEM. Chinese Chemical Letters. 2019; **30**(7):1401-1404. DOI: 10.1016/j.cclet.2019.02.024

[55] Chopra NG, Zettl A. Measurement of the elastic modulus of a multi-wall boron

nitride nanotube. Solid State Communications. 1998;**105**(5):297-300. DOI: 10.1016/S0038-1098(97)10125-9

[56] Suryavanshi AP, Yu MF, Wen J, Tang C, Bando Y. Elastic modulus and resonance behavior of boron nitride nano-tubes. Applied Physics Letters. 2004;**84**(14):2527-2529. DOI: 10.1063/1.1691189

[57] Zhou X, Tang DM, Mitome M, Bando Y, Sasaki T, Golberg D. Intrinsic and defect-related elastic moduli of boron nitride nanotubes as revealed by in situ transmission electron microscopy. Nano Letters. 2019;**19**(8):4974-4980. DOI: 10.1021/acs.nanolett.9b01170

[58] Golberg D, Costa PMFJ, Lourie O, Mitome M, Bai X, Kurashima K, et al. Direct force measurements and kinking under elastic deformation of individual multiwalled boron nitride nanotubes. Nano Letters. 2007;**7**(7):2146-2151. DOI: 10.1021/nl070863r

[59] Ghassemi HM, Lee CH, Yap YK, Yassar RS. Real-time fracture detection of individual boron nitride nanotubes in severe cyclic deformation processes. Journal of Applied Physics. 2010;**108**:024314. DOI: 10.1063/1.3456083

[60] Tanur AE, Wang J, Reddy AL, Lamont DN, Yap YK, Walker GC. Diameter-dependent bending modulus of individual multiwall boron nitride nanotubes. The Journal of Physical Chemistry. B. 2013;**117**(16):4618-4625. DOI: 10.1021/jp308893s

[61] Şahin H, Cahangirov S, Topsakal M, Bekaroglu E, Akturk E, Senger RT, et al. Monolayer honeycomb structures of group-IV elements and III-V binary compounds: First-principles calculations. Physical Review B. 2009;**80**(15):155453. DOI: 10.1103/PhysRevB.80.155453

[62] Kochaev A. Elastic properties of noncarbon nanotubes as compared to carbon nanotubes. Physical Review B. 2017;**96**(15):155428. DOI: 10.1103/PhysRevB.96.155428

[63] Menon M, Srivastava D. Structure of boron nitride nanotubes: Tube closing versus chirality. Chemical Physics Letters. 1999;**307**(5):407-412. DOI: 10.1016/S0009-2614(99)00552-7

[64] Jiang L, Guo W. A molecular mechanics study on size-dependent elastic properties of single-walled boron nitride nano-tubes. Journal of the Mechanics and Physics of Solids. 2011;**59**(6):1204-1213. DOI: 10.1016/j.jmps.2011.03.008

[65] Rajan AG, Strano MS, Blankschtein D. Ab initio molecular dynamics and lattice dynamics-based force field for modeling hexagonal boron nitride in mechanical and interfacial applications. The Journal of Physical Chemistry Letters. 2018;**9**(7):1584-1591. DOI: 10.1021/acs.jpclett.7b03443

[66] Mayo SL, Olafson BD, Goddard WA. DREIDING: A generic force field for molecular simulations. The Journal of Physical Chemistry. 1990;**94**(26):8897-8909. DOI: 10.1021/j100389a010

[67] Rappé AK, Casewit CJ, Colwell KS, Goddard WA III, Skiff WM. UFF, a full periodic table force field for molecular mechanics and molecular dynamics simulations. Journal of the American Chemical Society. 1992;**114**(25):10024-10039. DOI: 10.1021/ja00051a040

[68] Li C, Chou TW. A structural mechanics approach for the analysis of carbon nanotubes. International Journal of Solids and Structures. 2003;**40**(10):2487-2499. DOI: 10.1016/S0020-7683(03)00056-8

[69] Genoese A, Genoese A, Salerno G. Hexagonal boron nitride nanostructures: a nanoscale mechanical modeling. Journal of Mechanics of Materials and Structures. 2020;**15**(2):249-275. DOI: 10.2140/jomms.2020.15.249

[70] Ansari R, Rouhi S, Mirnezhad M, Aryayi M. Stability characteristics of single-walled boron nitride nanotubes. Archives of Civil and Mechanical Engineering. 2015;**15**(1):162-170. DOI: 10.1016/j.acme.2014.01.008

[71] Sakharova NA, Antunes JM, Pereira AFG, Chaparro BM, Fernandes JV. On the determination of elastic properties of single-walled boron nitride nanotubes by numerical simulation. Materials. 2021;**14**(12):3183. DOI: 10.3390/ma14123183

[72] Sakharova NA, Pereira AFG, Antunes JM, Fernandes JV. Numerical simulation study of the elastic properties of single-walled carbon nanotubes containing vacancy defects. Composites Part B: Engineering. 2016;**89**:155-168. DOI: 10.1016/j.compositesb.2015.11.029

[73] Chen Y, Chadderton LT, Gerald JF, Williams JS. A solid state process for formation of boron nitride nanotubes. Applied Physics Letters. 1999;**74**: 2960-2962. DOI: 10.1063/1.123979

[74] Kudin KN, Scuseria GE, Yakobson BI. C2F, BN, and C nanoshell elasticity from ab initio computations. Physical Review B. 2001;**64**(23):235406. DOI: 10.1103/PhysRevB.64.235406

[75] Zhang DB, Akatyeva E, Dumitrica T. Helical BN and ZnO nanotubes with intrinsic twisting: An objective molecular dynamics study. Physical Review B. 2011;**84**(11):115431. DOI: 10.1103/PhysRevB.84.115431

[76] Vijayaraghavan V, Zhang L. Consistent computational modeling of mechanical properties of carbon and boron nitride nanotubes. JOM. 2020;**72**: 3968-3976. DOI: 10.1007/s11837-020-04287-1

[77] Verma V, Jindal VK, Dharamvir K. Elastic moduli of a boron nitride nanotube. Nanotechnology. 2007; **18**(43):435711. DOI: 10.1088/0957-4484/18/43/435711

[78] Le MQ, Nguyen DT. Determination of elastic properties of hexagonal sheets by atomistic finite element method. Journal of Computational and Theoretical Nanoscience. 2015;**12**(4): 566-574. DOI: 10.1166/jctn.2015.3767